Introduction to Combinatorics

WILEY-INTERSCIENCE
SERIES IN DISCRETE MATHEMATICS AND OPTIMIZATION

ADVISORY EDITORS

RONALD L. GRAHAM
AT & T Bell Laboratories, Murray Hill, New Jersey, U.S.A.

JAN KAREL LENSTRA
Centre for Mathematics and Computer Scence, Amsterdam, The Netherlands
Erasmus University, Rotterdam, The Netherlands

ROBERT E. TARJAN
Princeton University, New Jersey, and
NEC Research Institute, Princeton, New Jersey, U.S.A.

AARTS AND KORST · Simulated Annealing and Boltzmann Machines: A Stochastic Approach to Combinatorial Optimization and Neural Computing

ALON, SPENCER, AND ERDÖS · The Probabilistic Method

ANDERSON AND NASH · Linear Programming in Infinite-Dimensional Spaces: Theory and Application

BARTHÉLEMY AND GUÉNOCHE · Trees and Proximity Representations

BAZARRA, JARVIS, AND SHERALI · Linear Programming and Network Flows

CHONG AND ZAK · An introduction to Optimization

COFFMAN AND LUEKER · Probabilistic Analysis of Packing and Partitioning Algorithms

DINITZ AND STINSON · Contemporary Design Theory: A Collection of Surveys

GLOVER, KLINGHAM, AND PHILLIPS · Network Models in Optimization and Their Practical Problems

GOLSHTEIN AND TRETYAKOV · Modified Lagrangians and Monotone maps in Optimization

GONDRAN AND MINOUX · Graphs and Algorithms (*Translated by S. Vajdā*)

GRAHAM, ROTHSCHILD, AND SPENCER · Ramsey Theory, Second Edition

GROSS AND TUCKER · Topological Graph Theory

HALL · Combinatorial Theory, Second Edition

JENSEN AND TOFT · Graph Coloring Problems

LAWLER, LENSTRA, RINNOOY KAN, AND SHMOYS, EDITORS · The Traveling Salesman Problem: A Guided Tour of Combinatorial Optimization

LEVITIN · Perturbation Theory in Mathematical Programming Applications

MAHMOUD · Evolution of Random Search Trees

MARTELLO AND TOTH · Knapsack Problems: Algorithms and Computer Implementations

MINC · Nonnegative Matrices

MINOUX · Mathematical Programming: Theory and Algorithms (*Translated by S. Vajdā*)

MIRCHANDANI AND FRANCIS, EDITORS · Discrete Location Theory

NEMHAUSER AND WOLSEY · Integer and Combinatorial Optimization

NEMIROVSKY AND YUDIN · Problem Complexity and Method Efficiency in Optimization (*Translated by E. R. Dawson*)

PACH AND AGARWAL · Combinatorial Geometry

PLESS · Introduction to the Theory of Error-Correcting Codes, Second Edition

SCHRIJVER · Theory of Linear and Integer Programming

TOMESCU · Problems in Combinatorics and Graph Theory (*Translated by R. A. Melter*)

TUCKER · Applied Combinatorics, Second Edition

MCALOON AND TRETKOFF · Optimization and Computational Logic

DASKIN · Network and Discrete Location: Modes, Algorithms and Applications

ERICKSON · Introduction to Combinatorics

Introduction to Combinatorics

Martin J. Erickson

A Wiley-Interscience Publication
JOHN WILEY & SONS, INC.
New York · Chichester · Brisbane · Toronto · Singapore

Library of Congress Cataloging in Publication Data:

Erickson, Martin J., 1963–
    Introduction to combinatorics / Martin J. Erickson.
        p.   cm. — (Wiley-Interscience series in discrete mathematics
    and optimization)
    "A Wiley-Interscience publication."
    Includes bibliographical references and index.
    ISBN 0-471-15408-3 (cloth : alk. paper)
    1. Combinatorial analysis.   I. Title.   II. Series.
QA164.E74   1996
511′.6—dc20   96-28174

Preface

Combinatorics may be described as the study of how discrete sets can be arranged, counted, and constructed. Accordingly, this book is an introduction to the three main branches of combinatorics: existence, enumeration, and construction. Within these three broad categories, combinatorics is concerned with many mathematical objects, including graphs, matroids, lattices, codes, designs, and algorithms. There are also several general theories, including Ramsey theory, the Pólya theory of counting, and the probabilistic method. No book can describe all the objects and theories, and some topics don't fit perfectly into one of the three aforementioned categories. There are overlaps and omissions. However, I believe that the three-part plan of this book is a good framework in which to introduce fundamental results, discuss interconnections and problem solving techniques, and collect open problems.

The format of this book is designed to gradually and systematically introduce most of the main concepts of combinatorics. Each part of the book opens with an easily stated "gem," then proceeds to develop a subject in depth, and finally arrives at a "big result." (The "big results" are Van der Waerden's theorem on arithmetic progressions, Pólya's graph enumeration formula, and Leech's 24-dimensional lattice.) In this way the reader is brought step-by-step from first principles to major accomplishments, always pausing to note mathematical points of interest along the way.

In order to illustrate the applicability of the combinatorial methods described in the book, I have paid careful attention to the selection of exercises at the end of each chapter. Many of the problems are quite challenging, and some are new. Included among the exercises are some problems from the William Lowell Putnam Mathematical Competition, an annual contest sponsored by the Mathematical Association of America and open to undergraduates in the United States and Canada. Although difficult, even the Putnam problems can be solved using the methods covered herein. To further reinforce understanding of the material, many important combinatorial methods, such as the pigeon-hole principle, are revisited several times throughout the book, i.e., in the exercises and examples as well as in the proofs of theorems.

This book evolved from lecture notes and problem sets used in teaching introductory combinatorics in the Division of Mathematics and Computer Science at Truman State University. I have found that there is enough material for a two-semester course, but a sampling of some sections from each of the three parts makes a reasonable one-semester course. In particular, Sections 2.4, 4.5, 4.6, 7.1–7.3, 8.5, 10.4–10.6, and 11.1–11.3 can be omitted without damaging the overall structure. The exposition of the book is at the advanced undergradute or beginning graduate student level. Although most of the discussion is self-contained, readers should possess certain prerequisite knowledge about discrete structures and algebraic structures. Specifically, one should be familiar with permutations and combinations, graphs and graph isomorphisms, and the basic abstract groups (cyclic, dihedral, symmetric, and alternating). Good references are Johnsonbaugh's *Discrete Mathematics* (1993) and Herstein's *Abstract Algebra* (1996). At any rate, the prerequisite material is presented briefly in Chapter 1.

Academic and research interest in combinatorics has grown tremendously since Leonhard Euler tackled the Königsberg bridge problem in 1736 and created graph theory. Today, there are more than twenty research journals in combinatorics and discrete mathematics, including *Journal of Combinatorial Theory Series A* and *Series B; Journal of Graph Theory; Discrete Mathematics; Discrete Applied Mathematics; Annals of Discrete Mathematics; Topics in Discrete Mathematics; SIAM Journal on Discrete Mathematics; Graphs and Combinatorics; Combinatorica; Ars Combinatoria; European Journal of Combinatorics; Journal of Algebraic Combinatorics; Journal of Combinatorial Designs; Designs, Codes and Cryptography; Journal of Combinatorial Mathematics and Combinatorial Computing; Combinatorics, Probability & Computing; Journal of Combinatorics, Information & System Sciences; Algorithms and Combinatorics; Random Structures & Algorithms; Bulletin of the Institute of Combinatorics and its Applications;* and *The Electronic Journal of Combinatorics* (on the World Wide Web). The connections between combinatorics and the theories of computing, information and codes, and probability are indicated by the titles of some of these journals. Indeed, it is now desirable for all mathematicians, statisticians, and computer scientists to be acquainted with the basic principles of combinatorics.

I wish to thank my three advisors from the University of Michigan for introducing me to combinatorics: Frank Harary, the world's pre-eminent graph theorist (or "G-man," as he says); Tom Storer, expert on cyclotomy; and Phil Hanlon, leader in algebraic combinatorics. I also wish to thank the many people who have kindly provided suggestions concerning this book: Robert Dobrow, Joe Flowers, Mark Schlatter, Michael Adam, Jan Karel Lenstra, Ronald Graham, Joel Spencer, Robert Tarjan, Dumont Hixson, Larry Wayne, Suren Fernando, Mark Faucette, Steve Smith, Bob Cacioppo, Rod Doll, Ke Tao, Todd Hammond, Ken Stilwell, Evan Haffner, Joe Hemmeter, David Dixon, Frank Sottile, Karen Singer, Olga Yiparaki, Carlos Montenegro, Milan Shah, George Ashline, Ken Plochinski, Dennis Chepurnov, Diann Reischman, Lata Potturi, H. Chad Lane, David Hardy, and Moreena Bond.

Finally, I wish to thank Michelle Correll and Cathleen Edmiston for their speedy and accurate word processing and Steve Quigley, Executive Editor of the Wiley-Interscience program, for his advice and support in completing the project.

M. E.

January 1996

Contents

Notation

Sets

$\mathcal{N}$	natural numbers
$\mathcal{Z}$	integers
$\mathcal{Q}$	rational numbers
$\mathcal{R}$	real numbers
$\mathcal{C}$	complex numbers
$\mathcal{N}_n = \{1, 2, \ldots, n\}$	set of the first n natural numbers
$P(A)$	power set of A
$[A]^t = \{B \subseteq A : |B| = t\}$	collection of t-subsets of A
$[n]^t = [\mathcal{N}_n]^t$	collection of t-subsets of the canonical n-set $\mathcal{N}_n$

Graphs

K_n	complete graph on n vertices
$K_{m,n}$	complete bipartite graph on m and n vertices
K_∞	complete graph on a countable set of vertices
$K_{\infty,\infty}$	complete bipartite graph on two countable sets of vertices
C_n	cycle with n vertices
P_n	path with n vertices

Groups

$\mathcal{Z}_n$	cyclic group of order n
D_n	dihedral group of order $2n$
S_n	symmetric group of order $n!$
A_n	alternating group of order $\frac{1}{2}n!$

1

Preliminaries: Set Theory, Algebra, and Number Theory

The purpose of Chapter 1 is to present the mathematical prerequisites necessary for understanding the rest of the book. This material consists of the most basic results in set theory, algebra, and number theory. As the reader has probably seen much or all of this material before, Chapter 1 could probably be skipped entirely, and if some background information is needed it could be looked up later. It should be pointed out that each topic in Chapter 1 is a large subject in its own right, and is only touched upon here.

1.1. SETS

As combinatorics deals primarily with relations and functions (on discrete sets), and relations and functions are defined in terms of sets, it makes sense to begin our review by talking about sets.

A *set* is a collection of *elements*. We sometimes write a set by listing its elements. For example,

$$A = \{1, 2, 3, 4, 5\}$$

is the set of integers between 1 and 5. We write $x \in A$ to indicate that x is an element of the set A. Thus $3 \in \{1, 2, 3, 4, 5\}$.

The *empty set*, denoted $\emptyset$, is the set with no elements.

We denote the set of *natural numbers* by $\mathcal{N}$; the set of *integers* (positive, negative, and 0) by $\mathcal{Z}$; the set of *rational numbers* by $\mathcal{Q}$; the set of *real numbers* by $\mathcal{R}$; and the set of *complex numbers* by $\mathcal{C}$. We denote the set $\{1, 2, \ldots, n\}$ by $\mathcal{N}_n$.

The *cardinality* of A, denoted $|A|$, is the number of elements of A. For example, $|\mathcal{N}_5| = 5$. If the cardinality of A is n, then we call A an *n-set*. Whenever A has only finitely many elements, we say that A is *finite*.

The *union* of two sets A and B, written $A \cup B$, is the collection of elements in A or B or both. The *intersection* of A and B, written $A \cap B$, is the collection of elements in both A and B.

We say that A is a *subset* of B, and write $A \subseteq B$, if every element of A is an element of B. The *power set* of A, denoted $P(A)$, is the collection of subsets of A. If the cardinality of A is n, then the cardinality of $P(A)$ is 2^n. (For each element of A, one must decide whether or not to include it in making a subset. Thus the number of subsets is equal to the number of outcomes of these n choices, i.e., 2^n.)

We let $[A]^t = \{B \subseteq A : |B| = t\}$. This set is called the collection of *t-subsets* of A. For convenience, we define $[n]^t = [\mathcal{N}_n]^t$.

1.2. RELATIONS AND FUNCTIONS

The *Cartesian product* $A \times B$ of two sets A and B is the collection of ordered pairs (a, b) with $a \in A$ and $b \in B$. If A and B are finite, with cardinalities x and y, respectively, then $A \times B$ has cardinality xy.

A *relation R* from A to B, written $R : A \longrightarrow B$, is a subset of $A \times B$. There are 2^{xy} possible relations from a set A of x elements to a set B of y elements, as this is the number of subsets of a set with xy elements. Note that if either A or B is empty, then of course $A \times B$ is empty, so the only relation is the empty set, and the formula 2^{xy} still applies. The *domain* of R is the set of $a \in A$ for which there exists $b \in B$ with $(a, b) \in R$; the *range* of R is the set of $b \in B$ for which there exists $a \in A$ with $(a, b) \in R$. If $(a, b) \in R$, then a is *related* to b. The *image* of a is $R(a) = \{b \in B : (a, b) \in R\}$, the set of elements to which a is related; the *preimage* of b is $R^{-1}(b) = \{a \in A : (a, b) \in R\}$, the set of elements related to b. The *converse relation* of R is $R^{-1} = \{(b, a) : (a, b) \in R\}$.

Figure 1.1 is a *directed graph* depiction of the relation $R : \{a, b, c\} \longrightarrow \{d, e\}$ with $R = \{(a, d), (a, e), (b, e), (c, e)\}$. The domain of R is $\{a, b, c\}$ and the range is $\{d, e\}$. Each ordered pair in R is represented by an arrow; e.g., (a, d) is represented by an arrow from a to d. In Exercise 1.1 the reader is asked to draw the directed graph representation of R^{-1}. Of course, it isn't necessary that the domain and range of a relation be distinct sets, as they are in our illustration.

A *relation R* on a set X is a relation $R : X \longrightarrow X$, from X to itself. For instance,

$$R_1 = \{(a, b) : a, b \in \mathcal{N}, a = b\}$$

and

$$R_2 = \{(a, b) : a, b \in \mathcal{N} \text{ and } b \text{ is divisible by } a\}$$

are two relations on the set $\mathcal{N}$ of natural numbers. These particular relations possess certain properties which we wish to abstract. A relation R on X is *reflexive* if $(a, a) \in R$ for all $a \in X$; *symmetric* if $(b, a) \in R$ whenever $(a, b) \in R$; *antisymmetric* if $(a, b) \in R$ and $(b, a) \in R$ imply $a = b$, for all $a, b \in X$; and *transitive* if $(a, b) \in R$ and $(b, c) \in R$ imply $(a, c) \in R$, for all $a, b, c \in X$. Corresponding to R_1 and R_2 above, we identify two types of relations of special interest.

1. An *equivalence relation* R on X is a reflexive, symmetric, transitive relation on X. The relation R_1 above is an equivalence relation on the set of natural numbers. If R is an equivalence relation on X, then, for each $a \in X$, the set $[a] = \{b \in X : (a, b) \in R\}$ is the *equivalence class* of a. An equivalence relation R on X is equivalent to what is called a partition of X. A *partition* P of X is a collection of nonempty disjoint subsets of X whose union equals X. Thus X is partitioned into equivalence classes of its elements, and, conversely, the elements of a partition of X are the equivalence classes of an equivalence relation on X. We will discuss the enumeration of partitions of a finite set in Chapters 5 and 6.

2. A *partial order* R on X is a reflexive, antisymmetric, transitive relation on X. The relation R_2 above is a partial order on $\mathcal{N}$. Often we denote a partial order by $\leq$, and write $a \leq b$ when $(a, b) \in R$. Two elements $a, b \in X$ are *comparable* if $a \leq b$ or $b \leq a$, and *incomparable* otherwise.

Equivalence relations and partial orders appear throughout mathematics, as the following sections will illustrate.

A *function* f from A to B, written $f : A \longrightarrow B$, is a relation from A to B for which the following condition holds: for each $a \in A$ there exists a unique $b \in B$ with $(a, b) \in f$. The set B is the *codomain* of f. We say that f *maps* a to b, and we write $f(a) = b$ or $f : a \mapsto b$. (The relation R in Figure 1.1 is not a function because a is related to two elements, d and e. However, a function can be created from R by removing (a, d) or (a, e).)

If A is empty, then the only function from A to B is the empty set; whereas if A is nonempty and B is empty, then there are no functions. Otherwise, if A has x elements and B has y elements, then there are y choices for where each of the x elements of A is mapped, so altogether there are y^x possible functions. Because every function is also a relation, the number of functions from A to B is less than or equal to the number

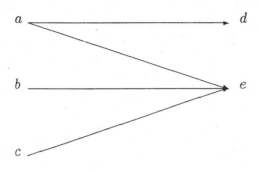

Figure 1.1. A relation with four elements.

of relations; that is, $y^x \leq 2^{xy}$ for all positive integers x and y. Observe that we have given a *counting* proof of an *algebraic* inequality. Counting proofs are highly desirable because they are often easier to "see" than other types of arguments. Alternatively, we can prove this inequality by mathematical induction or by taking logarithms base two, and the reader is asked to try these procedures in the exercises.

Let us assume that A and B are nonempty. The function $f : A \longrightarrow B$ is *one-to-one* (1–1) if no two elements of A are mapped to the same element of B. We say that f is *onto* if each $b \in B$ is the image of some $a \in A$. Equivalently, f is onto if the range of f equals B. If f is one-to-one and onto, then f is a *bijection*. If f is a bijection, then its converse relation f^{-1} is called the *inverse* of f. The bijections $f : A \longrightarrow B$ are the only functions whose converse relations $f^{-1} : B \longrightarrow A$ are also functions. Note that if f is a bijection, then $f(a) = b$ if and only if $f^{-1}(b) = a$.

Chapters 5 and 6 are concerned with the problems of counting one-to-one functions and onto functions from one finite set to another. In Section 2.1 we discuss the average number of elements in the sets $R(a)$ and $R^{-1}(b)$.

The following theorem is proved in most books on set theory. It is needed in the later sections on group theory and number theory.

Theorem 1.1. *Suppose f is a function from X to Y, where X and Y are two sets of the same finite cardinality. Then f is one-to-one if and only if f is onto.*

1.3. BINOMIAL COEFFICIENTS

How many one-to-one functions are possible from a set X of k elements into a set Y of n elements, assuming $n \geq k$? If we assign an order to the elements of X, then there are n choices for where to map the first element, followed by $n - 1$ choices for where to map the second element, etc. Thus the total number of one-to-one functions is $n(n-1)\cdots(n-k+1)$.

We let $n! = \prod_{k=1}^{n} k$ and $0! = 1$. From what was said above, the number of one-to-one functions from X into Y is $P(n,k) = n!/(n-k)!$. We call this expression *n permute k*.

If the elements of X are considered indistinct, then a one-to-one function f is equivalent to an unordered k-element subset of the n-set Y. To adjust for the unordering of the subset, we must divide $P(n,k)$ by $k!$. Thus the number of unordered subsets is given by the *binomial coefficient* $\binom{n}{k} = n!/[k!(n-k)!]$. We call this expression *n choose k*. The subsets are called *k-subsets* of Y. As there is a bijection between the k-subsets of Y and the $n-k$-subsets of Y, the binomial coefficients $\binom{n}{k}$ and $\binom{n}{n-k}$ are equal. This fact also follows from the formulas for the two expressions. The reason for the term "binomial coefficient" is found in the following theorem.

Theorem 1.2. *(Binomial theorem). For $a, b \in \mathcal{R}$ and $n \geq 0$,*

$$(a+b)^n = \sum_{k=0}^{n} \binom{n}{k} a^k b^{n-k}. \tag{1.1}$$

Proof. We give a combinatorial proof showing that the coefficients of $a^k b^{n-k}$ on the two sides of (1.1) are equal. On the left side the coefficient of $a^k b^{n-k}$ is the number of ways of selecting k factors $(a+b)$ which contribute a's (the other factors contribute b's). But this is the number of ways of selecting k unordered objects from a set of n objects, i.e., $\binom{n}{k}$. This agrees with the coefficient on the right side of (1.1). $\square$

Two combinatorial identities follow instantly from the binomial theorem. Putting $a = b = 1$ we obtain

$$2^n = \sum_{k=0}^{n} \binom{n}{k}. \tag{1.2}$$

Combinatorially, this identity says that the number of subsets of an n-set is the same as the number of ways of choosing a k-subset of an n-set, for any k.

Letting $a = -1$ and $b = 1$ we obtain

$$0 = \sum_{k=0}^{n} (-1)^k \binom{n}{k}. \tag{1.3}$$

This identity has an interesting combinatorial interpretation. We consider the equivalent formulation

$$\sum_{k \text{ odd}} \binom{n}{k} = \sum_{k \text{ even}} \binom{n}{k}, \tag{1.4}$$

which says that, for any n, the number of subsets of $X = \{1, \ldots, n\}$ with an odd number of elements is the same as the number of subsets with an even number of elements. For n odd, this assertion follows trivially from the symmetry of the binomial coefficients. We now give a combinatorial argument valid for any n. Let

$$A = \{S \subseteq X : |S| \text{ is even and } 1 \in S\}$$

$$B = \{S \subseteq X : |S| \text{ is odd and } 1 \in S\}$$

$$C = \{S \subseteq X : |S| \text{ is even and } 1 \notin S\}$$

$$D = \{S \subseteq X : |S| \text{ is odd and } 1 \notin S\}.$$

The obvious bijections between A and D and between B and C establish that $|A| = |D|$ and $|B| = |C|$. The identity (1.4) follows immediately.

This method of proving an identity by exhibiting two pairs of bijections is called the involution principle. See Stanton and White (1986).

Other binomial coefficient identities may be obtained by comparing like powers of x in certain algebraic identities. For example, comparing coefficients of x^k in the algebraic identity

$$(x + 1)^{m+n} = (x + 1)^m (x + 1)^n$$

we arrive at *Vandermonde's identity*

$$\binom{m + n}{k} = \sum_{i=0}^{k} \binom{m}{i} \binom{n}{k - i}. \tag{1.5}$$

This identity has a combinatorial interpretation. The binomial coefficient $\binom{m+n}{k}$ is the number of k-subsets of the $m + n$-set $A \cup B$, where $A = \{1, \ldots, m\}$ and $B = \{m + 1, \ldots, m + n\}$. The number of such subsets which contain i elements of A is $\binom{m}{i}\binom{n}{k-i}$. The summation $\sum_{i=0}^{k}\binom{m}{i}\binom{n}{k-i}$ counts these subsets for all i.

When $m = 1$, (1.5) becomes *Pascal's identity*

$$\binom{n+1}{k} = \binom{n}{k} + \binom{n}{k-1}. \tag{1.6}$$

Pascal's identity has a particularly simple combinatorial interpretation. The binomial coefficient $\binom{n+1}{k}$ is the number of k-subsets of the $n+1$-set $\{1,\ldots,n+1\}$. Each such subset contains or does not contain the element 1. The number of k-subsets which contain 1 is $\binom{n}{k-1}$. The number of k-subsets which do not contain 1 is $\binom{n}{k}$. The identity follows from this observation.

Putting $k = m = n$ in (1.5) we obtain another useful identity,

$$\binom{2n}{n} = \sum_{i=0}^{n}\binom{n}{i}^{2}. \tag{1.7}$$

This identity has an interesting combinatorial interpretation. The binomial coefficient $\binom{2n}{n}$ is the number of northeast paths which start at the southwest corner of an $n \times n$ grid and end at the northeast corner. Such paths have length $2n$ and are determined by a sequence of n "easts" and n "norths" in some order. The number of such sequences is $\binom{2n}{n}$. The summation $\sum_{i=0}^{n}\binom{n}{i}^{2}$ counts the paths according to their intersection with the main diagonal of the grid. That is, the number of paths which cross the diagonal at the point i units east of the starting point is $\binom{n}{i}^{2}$, for each i, where $0 \leq i \leq n$.

Here is another algebraic identity:

$$(x+1)^{m+n+1} = (x+1)\cdots(x+1), \tag{1.8}$$

where the right side has $m+n+1$ factors equal to $x+1$. The coefficient of x^{n+1} on the left side is $\binom{m+n+1}{n+1}$. On the right side, there is a contribution to x^{n+1} whenever we multiply x's from $n+1$ of the factors. Suppose the rightmost factor which contributes an x is the $n+i+1$-st factor, where $0 \leq i \leq m$. This leaves us the freedom to choose n other x's from a set of $n+i$ factors. Thus the coefficient of x^{n+1} on the right side of (1.8) is $\sum_{i=0}^{m}\binom{n+i}{n}$, proving the identity

$$\binom{m+n+1}{n+1} = \sum_{i=0}^{m}\binom{n+i}{n}. \tag{1.9}$$

This identity has a combinatorial interpretation. The binomial coefficient $\binom{m+n+1}{n+1}$ is the number of $n+1$-subsets of the $m+n+1$-set $\{1,\ldots,m+n+1\}$. Suppose the largest element in such a subset is $n+i+1$, where $0 \leq i \leq m$. The number of such subsets is $\binom{n+i}{n}$, and the summation $\sum_{i=0}^{m} \binom{n+i}{n}$ counts them all.

A change of variables in (1.9) yields the equivalent identity

$$\binom{n+1}{k+1} = \sum_{i=k}^{n} \binom{i}{k}. \tag{1.10}$$

It follows from (1.10) that

$$\binom{n+1}{k+1} - \binom{m+1}{k+1} = \sum_{i=m+1}^{n} \binom{i}{k}. \tag{1.11}$$

We conclude by proving a binomial coefficient identity used in proving Bonferroni's inequalities (Section 6.1). We start with the algebraic identity

$$(1+x)^{m-1} = (1+x)^{m}(1 - x + x^2 - x^3 + \cdots). \tag{1.12}$$

Equating coefficients of x^k on the two sides of this equation yields

$$(-1)^k \binom{m-1}{k} = \sum_{i=0}^{k} \binom{m}{i}(-1)^i. \tag{1.13}$$

Identity (1.13) can also be proved algebraically, by applying Pascal's identity to the binomial coefficient $\binom{m}{i}$ and then collapsing the resulting telescoping sum.

More identities are found in the exercises. The idea of comparing like powers of x in an algebraic identity is part of the great theory of generating functions, which we touch upon in Chapter 6.

1.4. GROUP THEORY

A *group* G is an ordered pair $(S, *)$ consisting of a nonempty set S and a binary operation $*$ on S, subject to the following three laws:

   1. For all $x, y, z \in S$, $x * (y * z) = (x * y) * z$ (associativity).

2. S contains an element e with the property that, for all $x \in S$, $x * e = e * x = x$. (e is called the *identity* element of G.)

3. For every $x \in S$ there exists an $x^{-1} \in S$ with the property that $x * x^{-1} = x^{-1} * x = e$. ($x^{-1}$ is called the *inverse* of x.)

One can easily prove that the identity element e of a group G is unique and that each element x has a unique inverse x^{-1}.

Abusing the notation, we often speak of G as though it contains the elements x, y, z, etc. Usually no confusion results. A *finite group* is one in which S is finite, and the *order* of a finite group is the number of elements in it. We often suppress the group operation sign and write xy for $x * y$. We abbreviate xx by x^2, $x^{-1}x^{-1}$ by x^{-2}, etc. For any $x \in G$, $x^0 = e$.

Example. $(\mathcal{Z}, +)$ is a group.

Example. The *cyclic group* $\mathcal{Z}_n$ of order n consists of the set $S = \{0, \ldots, n-1\}$ and the clock addition operation $*$ defined by $x * y = x + y$ if $x + y < n$ and $x * y = x + y - n$ if $x + y \geq n$.

A group G is *abelian* if $xy = yx$ for all $x, y \in G$. Otherwise, G is *nonabelian*.

Example. $\mathcal{Z}_n$ is an abelian group.

The *order* of an element $x \in G$ is the least positive integer n for which $x^n = e$. If there is no such integer, then we say that x has *infinite order*.

Example. In $\mathcal{Z}_4$, the elements 0, 1, 2, 3 have orders 1, 4, 2, 4, respectively.

Example. The *symmetric group* S_n of order $n!$ consists of the set of bijections f from $\mathcal{N}_n$ to $\mathcal{N}_n$. The group operation $*$ is composition of bijections. The elements of S_n are conveniently written in *cycle notation*. Thus

$$(1\ 2\ 3)(4\ 8)(3\ 6\ 7)(9)$$

is the permutation element of S_9 which sends 1 to 2 to 3 to 1, *transposes* 4 and 8, sends 3 to 6 to 7 to 3, and *fixes* 9. To multiply two permutations together, we just compute the result of the composition of the two bijections (reading left to right). For example,

$$(1\ 2\ 3)(4\ 5) * (1\ 2\ 3\ 4\ 5) = (1\ 3\ 2\ 4)(5).$$

Because $(1\ 2)(1\ 3) \neq (1\ 3)(1\ 2)$, it follows that S_n is nonabelian for $n \geq 3$.

A *homomorphism* from one group $G_1 = (S_1, *_1)$ to a second group $G_2 = (S_2, *_2)$ is a map f which preserves multiplicative structure: $f(x *_1 y) = f(x) *_2 f(y)$ for all $x, y \in S_1$. If the homomorphism is a bijection we call it an *isomorphism* and say that the two groups G_1 and G_2 are *isomorphic*; we write $G_1 \simeq G_2$. A one-to-one homomorphism is called a *monomorphism* and an onto homomorphism is called an *epimorphism*. An isomorphism from a group G to itself is called an *automorphism* of G.

Suppose G_1 and G_2 are two groups. We define their *product* $G_1 \times G_2$ to be the set of ordered pairs $\{(g_1, g_2) : g_1 \in G_1, g_2 \in G_2\}$ subject to the multiplication rule $(g_1, g_2) * (g_3, g_4) = (g_1 g_3, g_2 g_4)$.

Example. $\mathcal{Z}_2 \times \mathcal{Z}_2$ is a four-element group. (It is not isomorphic to $\mathcal{Z}_4$, for $\mathcal{Z}_2 \times \mathcal{Z}_2$ has three elements of order 2 and $\mathcal{Z}_4$ has only one.)

We say that H is a *subgroup* of G if H is a subset of G and H is a group with respect to the group operation for G.

Example.   The two-element group $\{(1\ 2)(3), (1)(2)(3)\}$ is a subgroup of the six-element group S_3.

The symmetric group S_n is especially important for it happens that every finite group is isomorphic to a subgroup of some S_n. This is Cayley's theorem.

Theorem 1.3.   *(Cayley's theorem). If G is a finite group of order n, then G is isomorphic to a subgroup of S_n.*

Proof.   For each element $x \in G$, we define a function f_x from G to G by the rule $f_x(a) = ax$ (right multiplication by x). Note that f_x is one-to-one, because $f_x(a) = f_x(b)$ implies $ax = bx$ which implies $a = b$. Because f is a one-to-one function from a finite set into itself, it follows that f is a bijection. Thus each f_x is a permutation of the n-element set G. The collection of these permutations is isomorphic to a subgroup of S_n. $\square$

Example.   The *dihedral group* D_n of order $2n$ consists of the set of symmetries of a regular n-gon. If we number the vertices of the n-gon

$1, \ldots, n$, then we see that D_n is a subgroup of S_n. Specifically, the subgroup is generated by two permutations: the *rotation* $r = (1\ 2\ 3\ 4\ \ldots\ n)$ and a *flip* f along an axis of symmetry of the n-gon. If n is odd we take the flip to be $f = (n)(1\ n-1)(2\ n-2)(3\ n-3) \ldots (\frac{n-1}{2}\ \frac{n+1}{2})$. If n is even we choose $f = (1\ n-1)(2\ n-2) \ldots (\frac{n}{2}-1\ \frac{n}{2}+1)$. Every element of D_n can be written in the form $r^a f^b$, $0 \leq a \leq n-1$, $0 \leq b \leq 1$. Two such elements are multiplied using the basic rules $r^n = e$, $f^2 = e$, and $rf = fr^{-1}$. From these basic rules it is possible to compute all other products. We say that D_n has the *presentation* $< r, f : r^n = 1, f^2 = 1, rf = fr^{-1} >$. For information about group presentations the reader may consult Johnson (1990).

We have noted that every element of S_n may be expressed as a product of cycles. A cycle of length 2 is called a *transposition* and a cycle of length 1 is called a *fixed point*. Cycles of length greater than 2 can be written as products of transpositions. For example, $(1\ 2\ 3) = (1\ 2)(1\ 3)$. It turns out that when a permutation is written as a product of transpositions and fixed points, the number of transpositions is always even or always odd. We refer the reader to Herstein (1996). The permutation is accordingly called an *even* permutation or an *odd* permutation. It follows from the first homomorphism theorem for groups, to be discussed shortly, that S_n contains $n!/2$ even permutations and $n!/2$ odd permutations. (This fact follows more simply from the observation that $f(\sigma) = (12)\sigma$ is a bijection between the set of even permutations of S_n and the set of odd permutations of S_n.) As the identity permutation is even and the even permutations are closed under multiplication and taking inverses, the even permutations are a group.

Example. The *alternating group* A_n of order $\frac{1}{2}n!$ consists of the even permutations of S_n.

Suppose H is a subgroup of G. A *left coset* of H in G is a set $xH = \{xh : h \in H\}$ for some $x \in G$. Note that each left coset of H has exactly the same number of elements as H. We define an equivalence relation R on G by letting $(x, y) \in R$ if and only if x and y are in the same left coset of H. We claim that R is an equivalence relation. The reflexivity property and symmetry property are trivially true. As for transitivity, if $(x, y) \in R$ and $(y, z) \in R$, then $x = rh$, $y = rh'$, $y = sh''$, and $z = sh'''$ for some $h, h', h'', h''' \in H$ and $r, s \in G$. Hence $z = rh'h''^{-1}h''' \in rH$, so that $(x, z) \in R$. Thus R is transitive and therefore is an equivalence relation. The number of equivalence classes is the *index* of H in G, written $[G : H]$. As G is partitioned into equivalence classes, we have shown that $|G| = |H|[G : H]$, from which Lagrange's theorem follows instantly.

Theorem 1.4. *(Lagrange's theorem). If G is a finite group and H is a subgroup of G, then the order of H divides the order of G.*

The converse of Lagrange's theorem is false. The smallest counterexample is A_4, a 12-element group containing no subgroup of order 6.

Corollary 1.5. *If G is a finite group of order n and x is an element of G, then $x^n = e$.*

Proof. Let $H = \{x^a : a \in \mathcal{Z}\}$. Because H is closed with respect to products and inverses, H is a (cyclic) subgoup of G. Therefore, by Lagrange's theorem, the order of H divides n. But the order of H is the minimum positive a for which $x^a = e$. Thus $n = ak$ for some k, and $x^n = x^{ak} = (x^a)^k = e$. $\square$

A *right coset* of H in G is a set $Hx = \{hx : h \in H\}$. If the collection of right cosets of H in G is the same as the collection of left cosets of H in G, then we say that H is a *normal subgroup* of G, and we write $H \lhd G$. If $H \lhd G$, then we define the *quotient group* G/H of G by H to be the set of left cosets xH of H under the multiplication rule $xH * yH = (xy)H$. Because H is a normal subgroup of G, this multiplication is well defined.

Suppose f is a homomorphism from G_1 to G_2, and suppose e_2 is the identity element of G_2. The preimage $f^{-1}(e_2)$ of e_2 is called the *kernel* of f. It turns out that the kernel K is a normal subgroup of G_1. We refer the reader to Herstein (1996).

Theorem 1.6. *(First homomorphism theorem for groups). Let f be a homomorphism from G_1 onto G_2, and suppose K is the kernel of f. Then G_2 is isomorphic to G_1/K.*

We refer the reader to any book on algebraic structures for a proof of this theorem.

The first homomorphism theorem for groups gives some further information about the groups $\mathcal{Z}_n$ and A_n. Let $n\mathcal{Z}$ be the group of integer multiples of n under addition. Let the homomorphism $f : \mathcal{Z} \longrightarrow n\mathcal{Z}$ be defined by $f(x) = y$, where $y \in \mathcal{Z}_n$ and $x \equiv y \bmod n$. Then the kernel of f is $n\mathcal{Z}$, and, by Theorem 1.6, $\mathcal{Z}_n$ is isomorphic to $\mathcal{Z}/n\mathcal{Z}$.

Similarly, we could have defined the alternating group A_n to be the kernel of the homomorphism $f : S_n \longrightarrow \{-1, 1\}$ which sends odd permutations to -1 and even permutations to $+1$. In particular, this shows that $A_n \simeq S_n/Z_2$ and $|A_n| = \frac{1}{2}n!$.

1.5. NUMBER THEORY

Number theory is concerned with divisibility properties of integers. For any integer $n \geq 2$, we say that two integers a and b are *congruent modulo n* if n divides $a - b$; and we write

$$a \equiv b \bmod n.$$

Because $n\mathcal{Z}$ is a (normal) subgroup of $\mathcal{Z}$, the modulo relation is the same as the relation introduced in the proof of Lagrange's theorem. That is, $a \equiv b \bmod n$ if and only if a and b are in the same coset of $n\mathcal{Z}$. We have remarked that such a relation is an equivalence relation. The equivalence classes are called *residue classes* (modulo n). They form a group which we have already denoted $\mathcal{Z}_n$. Recall that the operation in $\mathcal{Z}_n$ is modular arithmetic (clock arithmetic). Now let us consider $\mathcal{Z}_n - \{0\}$, the set of nonzero residue classes in $\mathcal{Z}_n$. When is $\mathcal{Z}_n - \{0\}$ a group under multiplication?

Theorem 1.7. $\mathcal{Z}_n - \{0\}$ *is a group under multiplication if and only if n is prime.*

Proof. Suppose that n is composite, $n = ab$ with $a, b > 1$. It follows that neither a nor b has a multiplicative inverse. For example, if a^{-1} exists, then $a^{-1}a \equiv 1 \bmod n$, which implies $b \equiv a^{-1}ab \equiv a^{-1}n \equiv 0 \bmod n$, a contradiction. Therefore $\mathcal{Z}_n - \{0\}$ is not a group if n is composite.

Now suppose n is prime and $a \in \mathcal{Z}_n - \{0\}$. We want to show that there is a multiplicative inverse for a. Define a function $f : \mathcal{Z}_n - \{0\} \longrightarrow \mathcal{Z}_n - \{0\}$ by the rule $f(x) = ax$. Note that the function is one-to-one, for if $ax \equiv ay \bmod n$, then n divides $a(x - y)$, and because n is relatively prime to a it follows that n divides $x - y$ and hence $x \equiv y \bmod n$. Because f is a one-to-one function between two sets of the same cardinality, f must be a bijection. Hence f is onto and there exists an x for which $f(x) \equiv ax \equiv 1 \bmod n$. This x is a multiplicative inverse for a. Thus $\mathcal{Z}_n - \{0\}$ is a multiplicative group. $\square$

We have determined that, for p prime, $\mathcal{Z}_p - \{0\}$ is a multiplicative group with $p - 1$ elements. Thus, by the corollary of Lagrange's theorem in the previous section, $x^{p-1} \equiv 1 \bmod p$ for any group element x. This is known as Fermat's little theorem.

Theorem 1.8.   *(Fermat's little theorem). If $x \not\equiv 0 \bmod p$, then*

$$x^{p-1} \equiv 1 \bmod p.$$

For any n, let $\mathcal{Z}_n^*$ be the set of elements relatively prime to n. We have proved that $\mathcal{Z}_p^*$ is a multiplicative group for p prime. By reasoning similar to that in the proof of Theorem 1.7, it turns out that $\mathcal{Z}_n^*$ is always a group. What is the order of this group? To answer this question, we introduce Euler's ϕ-function, defined for all natural numbers as follows:

$$\phi(n) = |\{x : 1 \leq x \leq n \text{ and } \gcd(x,n) \doteq 1\}|.$$

Thus $\phi(n)$ is the order of the group $\mathcal{Z}_n^*$. In Chapter 6 we establish the following formula for $\phi(n)$:

$$\phi(n) = n \prod_{i=1}^{k} \left(1 - \frac{1}{p_i}\right),$$

where the canonical factorization of n into prime powers is $n = \prod_{i=1}^{k} p_i^{\alpha_i}$.

The following theorem of L. Euler is a generalization of Fermat's little theorem.

Theorem 1.9.   *(Euler's theorem). If $\gcd(x,n) = 1$, then*

$$x^{\phi(n)} \equiv 1 \bmod n.$$

Euler's theorem allows us to solve linear congruences. The linear congruence

$$ax \equiv b \ (\bmod\ n),$$

where $\gcd(a,n) = 1$, has the solution

$$x \equiv a^{\phi(n)-1}b \ (\bmod\ n).$$

However, when we consider congruences of the form

$$x^2 \equiv a \ (\bmod\ n),$$

it isn't immediately clear whether solutions exist. That depends on the value of a.

The $p - 1$ nonzero residues of $\mathcal{Z}_p$ fall into two classes: quadratic residues and quadratic nonresidues. If $y \equiv x^2$ modulo p for some nonzero x, then y is a *quadratic residue (modulo p)*; if not, then y is a *quadratic nonresidue (modulo p)*. Thus the set of quadratic residues modulo p is

$$QR = \{x^2 : x \in \mathcal{Z}_p^*\}$$

and the set of quadratic nonresidues is

$$QN = \mathcal{Z}_p^* - QR.$$

It is helpful to recognize QR as the range of the homomorphism $f : \mathcal{Z}_p^* \longrightarrow \mathcal{Z}_p^*$, $f(x) = x^2$. If p is any prime greater than 2, then the map $f : \mathcal{Z}_p^* \longrightarrow QR$, $f(x) = x^2$ is a homomorphism with kernel $\{-1, 1\}$. It follows from the first homomorphism theorem for groups that $|QR| = (p - 1)/2$, and therefore also that $|QN| = (p - 1)/2$. We define the *Legendre symbol* (x/p) as follows:

$$\left(\frac{x}{p}\right) = 0 \text{ if } x \equiv 0 \bmod p \tag{1}$$

$$= 1 \text{ if } x \in QR \tag{2}$$
$$= -1 \text{ if } x \in QN.$$

By the above considerations, $\sum(x/p) = 0$ for any sum over a complete residue system modulo p.

Note. The Legendre symbol (x/p) is also called the *quadratic character* of x *(modulo p)*, written $\chi(x)$.

1.6. FIELDS

A *field F* is an ordered triple $(S, +, \cdot)$ consisting of a nonempty set S and two binary operations $+$ and $\cdot$ such that $(S, +)$ is an abelian group with identity 0, $(S - \{0\}, \cdot)$ is an abelian group with identity 1, and the distributive law $x \cdot (y + z) = x \cdot y + x \cdot z$ holds for all $x, y, z \in S$. We also require that $0 \neq 1$, or, equivalently, that S has at least two elements. Just as with groups, we usually consider F as containing the elements of S.

If we drop the requirement that each element has a multiplicative inverse, then the resulting structure is called a *commutative ring with identity*. In this book we deal mainly with groups and fields. However, the binomial theorem (proved earlier) is true for commutative rings with identity.

Example. The familiar structures $\mathcal{R}$ and $\mathcal{C}$ are fields.

A *finite* field is a field with a finite number of elements.

Example. We proved in the previous section that $\mathcal{Z}_p$ is a finite field whenever p is prime.

One can easily prove that for any finite field F there is a smallest positive integer p (called the *characteristic* of the field) with the property that $px = 0$ for all $x \in F$. It follows that p is prime and that the field has p^k elements for some k. This is a consequence of the fact that F is a vector space over $\mathcal{Z}_p$. (We discuss vector spaces in the next section.) Furthermore, one can show that there is a unique field of order p^k up to isomorphism. We call this field the *Galois field* $GF(p^k)$, named after E. Galois, the discoverer of the connection between groups and algebraic equations. It can be proved that the nonzero elements of $GF(p^k)$ form a cyclic group under multiplication. For proofs of these facts see Herstein (1996).

1.7. LINEAR ALGEBRA

A *vector space* V over a field F is an ordered 4-tuple $(S, F, +, \cdot)$ consisting of a nonempty set S of *vectors*, a field F, a map $+$ from $S \times S$ to S such that $(S, +)$ is an abelian group, and a map $\cdot$ from $F \times S$ to S, such that the following conditions hold for all $f, f_1, f_2 \in F$, $v, v_1, v_2 \in S$:

   1. $f \cdot (v_1 + v_2) = f \cdot v_1 + f \cdot v_2$.
   2. $(f_1 + f_2) \cdot v = f_1 \cdot v + f_2 \cdot v$.
   3. $f_1 \cdot (f_2 v) = (f_1 f_2) \cdot v$.
   4. $1 \cdot v = v$, where 1 is the multiplicative identity of F.

As expected, we usually speak of V as containing the elements of S and we write $f \cdot v$ as fv.

An important example of a vector space is F^n, the vector space of sequences of n elements of the field F. We usually write an element of F^n

as an $n \times 1$ *vector*. For example, with $n = 4$ and $F = GF(2)$, a typical element is

Suppose that V is a vector space and S is a subset of V. If every element $v \in V$ can be written as a linear combination of elements of S over F, i.e.,

$$v = f_1 v_1 + \cdots + f_n v_n$$

for some elements $v_1, \ldots, v_n$ of S and elements $f_1, \ldots, f_n$ of F, then we say that S *spans* V (*over F*).

If no element of S can be written as a linear combination of the other elements of S, then we say that S is *linearly independent* (*over F*).

If S is linearly independent over F and S spans V over F, then we say that S is a *basis* for V (over F). The following theorem is proved in most books on linear algebra.

Theorem 1.10. *If V is a vector space over F, then V has a basis (over F). Moreover, all bases of V over F have the same cardinality.*

The cardinality of a basis is called the *dimension* of the vector space V (*over F*).

In linear algebra we are interested in linear transformations from one vector space to another. For example, a linear transformation from F^3 to F^3 can be written

$$y_1 = a_{11} x_1 + a_{12} x_2 + a_{13} x_3$$
$$y_2 = a_{21} x_1 + a_{22} x_2 + a_{23} x_3$$
$$y_3 = a_{31} x_1 + a_{32} x_2 + a_{33} x_3$$

where the coefficients a_{ij} are elements of F. This transformation is written in matrix form as

$$\begin{bmatrix} y_1 \\ y_2 \\ y_3 \end{bmatrix} = \begin{bmatrix} a_{11} & a_{12} & a_{13} \\ a_{21} & a_{22} & a_{23} \\ a_{31} & a_{32} & a_{33} \end{bmatrix} \begin{bmatrix} x_1 \\ x_2 \\ x_3 \end{bmatrix}.$$

Matrix addition is defined componentwise. Thus the sum of matrices $A = [a_{ij}]_{p \times q}$ and $B = [b_{ij}]_{p \times q}$ is $A + B = [a_{ij} + b_{ij}]_{p \times q}$. The identity matrix is the all zero matrix of the same dimensions.

Matrix multiplication is defined so that the matrix associated with the composition of two linear transformations is the product of the two matrices. Specifically, the matrix product of $A = [a_{ij}]_{p \times q}$ and $B = [b_{ij}]_{q \times r}$ is $AB = [c_{ij}]_{p \times r}$, where $c_{ij} = \sum_k a_{ik} b_{kj}$. For example,

$$\begin{bmatrix} a & b \\ c & d \end{bmatrix} \begin{bmatrix} A & B \\ C & D \end{bmatrix} = \begin{bmatrix} aA + bC & aB + bD \\ cA + dC & cB + dD \end{bmatrix}.$$

The identity matrix with respect to matrix mutliplication is $I_n = [a_{ij}]_{n \times n}$, where $a_{ij} = 1$ if $i = j$ and $a_{ij} = 0$ otherwise.

It is of interest to know whether a matrix transformation has an inverse. It turns out that a square matrix has an inverse if and only if a quantity called its *determinant* is nonzero. If

$$A = [a_{ij}]$$

is an $n \times n$ matrix, then the determinant of A is given by the formula

$$\det A = \sum_{\pi} sgn(\pi) a_{1\pi(1)} a_{2\pi(2)} \cdots a_{n\pi(n)}$$

where the sum is taken over all permutations π of $\{1, \ldots, n\}$, and $sgn(\pi)$ is $+1$ if π is even and -1 if π is odd. This amounts to summing all signed products of "nonattacking rook patterns" of entries of A.

If A^{-1} exists, then the transformation given by A is one-to-one: $Ax = Ay$ implies $A(x - y) = 0$, which implies $A^{-1}A(x - y) = A^{-1}0 = 0$. This in turns implies $x - y = 0$, or $x = y$.

The following theorems are proved in most books on matrix algebra or linear algebra.

Theorem 1.11. *The transformation given by a square matrix A has an inverse if and only if $\det(A)$ is nonzero.*

Theorem 1.12. *For any two square matrices of the same size,*

$$\det(AB) = \det(A)\det(B).$$

Let A be an $m \times n$ matrix and consider the function f from $\mathcal{R}^n$ to $\mathcal{R}^m$ defined by sending vector v to vector Av. Observe that f is a homomorphism between the additive groups $\mathcal{R}^n$ and $\mathcal{R}^m$. The kernel of this homomorphism is called the *null space* of A. The dimension of the kernel is called the *nullity* of A, written $n(A)$. The range of f is called the *range space* of A. The dimension of the range space is called the *rank* of A, written $r(A)$. The following theorem is proved in most books on linear algebra.

Theorem 1.13.   *If A is an $m \times n$ matrix, then $r(A) + n(A) = n$.*

Example.   The matrix

$$\begin{bmatrix} 1 & 2 \\ 2 & 4 \end{bmatrix}$$

has rank 1, as it sends every vector

$$\begin{bmatrix} x \\ y \end{bmatrix}$$

onto a 1-dimensional subspace of $\mathcal{R}^2$, namely, the line $y = 2x$. Theorem 1.13 says that the nullity of this matrix is 1, and we observe that the kernel is the set of vectors on the line $y = -\frac{1}{2}x$. Thus $r(A) + n(A) = 1 + 1 = 2$.

Notes

The reader may wish to consult Herstein (1996) for a good introduction to abstract algebra. A good reference for number theory is Niven, Zuckerman, and Montgomery (1991).

Some problems in the exercise sets have appeared as questions on the William Lowell Putnam Mathematical Competition, an annual contest sponsored by the Mathematical Association of America and open to undergraduates in the United States and Canada. These problems, though challenging, can be solved by the methods described in the various chapters of this book. In a few cases, we have supplied hints which were not given in the problems as originally stated in the contests.

Exercises

1.1 Draw the inverse relation R^{-1} of the relation R in Figure 1.1.

1.2 Prove that $y^x \leq 2^{xy}$ for all positive integers x and y by using mathematical induction or by taking logarithms base two of both sides of the inequality.

1.3 Furnish examples of functions from $\mathcal{N}$ to $\mathcal{N}$ which are (1) one-to-one and onto, (2) one-to-one and not onto, (3) onto and not one-to-one, and (4) not one-to-one and not onto.

1.4 Give an example of a bijection from $\mathcal{N}$ to $\mathcal{N} \cup \{0\}$.

1.5 Give an example of a bijection from the closed interval $[0,1]$ to the open interval $(0,1)$. Hint: Let f be the identity function except for $f(0) = \frac{1}{3}, f(\frac{1}{3^a}) = \frac{1}{3^{a+1}}, f(1) = \frac{1}{2}$, and $f(\frac{1}{2^b}) = \frac{1}{2^{b+1}}$.

1.6 (Tarski's fixed point theorem) Let S be a set and let $f : P(S) \longrightarrow P(S)$ be a function for which $A \subseteq B$ implies $f(A) \subseteq f(B)$. Show that there exists a set $A \in P(S)$ with $f(A) = A$.   Hint: Let $C = \{B \in P(S) : B \subseteq f(B)\}$ and $A = \bigcup_{c \in C} c$. Show that $A \subseteq f(A)$ and $f(A) \subseteq A$.

1.7 (Putnam Competition, 1955) On a circle, n points are selected and the chords joining them in pairs are drawn. Assuming that no three of these chords are concurrent (except at the endpoints), how many points of intersection are there?

1.8 (Putnam Competition, 1959) Let each of m distinct points on the positive part of the X-axis be joined to n distinct points on the positive part of the Y-axis. Obtain a formula for the number of intersection points of these segments (exclusive of endpoints), assuming that no three of the segments are concurrent.

1.9 Prove the identity $\binom{n}{k}\binom{k}{j} = \binom{n}{j}\binom{n-j}{k-j}$. This relation is called the "subcommittee identity" because both sides count the number of ways to choose, from n people, a committee of size k and a subcommittee of size j.

1.10 Suppose five particles are traveling back and forth on the unit interval $I = [0, 1]$. Initially, all five particles move to the right with the same speed. (The initial placement of the particles is immaterial, as long as they are not at the endpoints.) When a particle reaches 0 or 1, it reverses direction but maintains its speed. When two particles collide, they both reverse direction (and maintain speeds). How many particle–particle collisions occur before the particles once again occupy their original positions and are moving to the right?

1.11 Show that the number of ways $2n$ people may be paired into n pairs is $\binom{2n}{n} n! 2^{-n}$.

1.12 Show that $(kn)!$ is divisible by $(n!)^k$. Hint: Think combinatorially!

1.13 Prove the identity $\binom{2n-1}{n} = \sum_{k=0}^{n} \binom{n}{k} \binom{n-1}{k}$.

1.14 Prove the identity $\sum_{i=0}^{n} \binom{n+i}{n} 2^{-i} = 2^n$.

1.15 Prove the identity $\binom{-a}{n} = \binom{a+n-1}{n}(-1)^n$.

1.16 Prove the identity $\sum_{i=0}^{n} \binom{n}{i}\binom{2n}{i} = \binom{3n}{n}$.

1.17 Prove that the identity element e of a group G is unique. Prove that each element x in G has a unique inverse x^{-1}

1.18 List the elements of S_4 which have order 2.

1.19 How many elements of S_{17} have order 210?

1.20 How many elements of $\mathcal{Z}_{100}$ have order 40?

1.21 Assume that G is a finite group of order n and a is a positive integer relatively prime to n. Show that the map $f : G \longrightarrow G$ defined by $f(x) = x^a$ is onto.

1.22 Illustrate Caley's theorem for the group $\mathcal{Z}_2 \times \mathcal{Z}_2$.

1.23 The *center* of G, denoted $C(G)$, is the set of elements of G which commute with all elements of G. Show that $C(G)$ is a normal subgroup of G. Show that $G/C(G)$ is cyclic if and only if G is abelian.

1.24 The group of *inner automorphisms* of G, denoted $\text{Inn}(G)$, is the group of automorphisms $f : G \longrightarrow G$ defined by $f(g) = xgx^{-1}$ for some $x \in G$. Show that $\mathcal{Z}/C(G)$ is isomorphic to $\text{Inn}(G)$.

1.25 Show that if $[G : H] = 2$, then $H \triangleleft G$.

1.26 Let $G = (\mathcal{R} - \{0\}, \cdot)$ and $N = (\{1, -1\}, \cdot)$. Prove $N \triangleleft G$ and $G/N \simeq (\mathcal{R}^+, \cdot)$.

1.27 Let G be an abelian group and let T consist of the elements of G of finite order. Show that T is a normal subgroup of G. T is called the *torsion* subgroup of G. Show that the nonidentity elements of G/T have infinite order.

1.28 Prove $\sum_{d|n} \phi(d) = n$.

1.29 What is the least positive residue modulo 47 which satisfies $7x \equiv 20 \bmod 47$?

1.30 What cyclic group or product of cyclic groups is isomorphic to $\mathcal{Z}_{18}^*$? What about $\mathcal{Z}_{98}^*$?

1.31 Show that the axiom that a field has at least two elements is equivalent to the axiom that $0 \neq 1$.

1.32 Let

$$A = \begin{bmatrix} 5 & 1 \\ 4 & 1 \end{bmatrix}.$$

Find $\det A$ and A^{-1}.

1.33 For any square matrix $A = [a_{ij}]_{n \times n}$, and any entry a_{ij} of A, prove that $\det A = a_{ij}X + Y$ for some numbers X and Y which do not depend on a_{ij}. See Section 10.6.

PART

Existence

I

Every sequence of ten distinct integers contains an increasing subsequence of four integers or a decreasing subsequence of four integers (or both). For example, the sequence $5, 8, -1, 0, 2, -4, -2, 1, 7, 6$ contains the increasing subsequence $-1, 0, 2, 7$.

This proposition is an existence result. No matter which ten integers are chosen, and no matter in what order they occur, there exists a specific type of subsequence (namely, a monotonic subsequence of four integers).

Part I of this book deals with existence theorems, from the most elementary (the pigeonhole principle), through beautiful gems like the one above (an instance of the Erdös–Szekeres theorem), to two key results of Ramsey theory (Ramsey's theorem and Van der Waerden's theorem). The central pursuit is always to find ordered subsets of large disordered systems. Or, put more poetically, we want to find order in randomness.

2

The Pigeonhole Principle

In this chapter we prove several versions of the pigeonhole principle, the simplest existence result in combinatorics, and give many applications.

2.1. VERSIONS OF THE PIGEONHOLE PRINCIPLE

The most important theorem in existential combinatorics is also the simplest: the pigeonhole principle. It occurs in many variations, a few of which we discuss here, and says, basically, that not every element in a set is below average and not every element is above average. We now state and prove a fairly general version.

Theorem 2.1. *(Pigeonhole principle). Let R be a relation from A to B, where A and B are finite nonempty sets. The following four statements hold:*

(1) There exists $a \in A$ with $|R(a)| \geq |R|/|A|$.

(2) There exists $a \in A$ with $|R(a)| \leq |R|/|A|$.

(3) There exists $b \in B$ with $|R^{-1}(b)| \geq |R|/|B|$.

(4) There exists $b \in B$ with $|R^{-1}(b)| \leq |R|/|B|$.

   Proof.   We prove (1) by contradiction. Assume $|R(a)| < |R|/|A|$ for all $a \in A$. Then

$$|R| = \sum_{a \in A} |R(a)|$$

$$< \sum_{a \in A} \frac{|R|}{|A|}$$

$$= |A| \cdot \frac{|R|}{|A|}$$

$$= |R|.$$

We conclude that $|R| < |R|$, an absurdity. Therefore our assumption that $|R(a)| < |R|/|A|$ for all a is false. Hence there is an $a \in A$ with $|R(a)| \geq |R|/|A|$.

We prove (2) by replacing '<' by '>' in the above argument. Finally, (3) and (4) follow from (1) and (2) by considering the converse relation $R^{-1} : B \longrightarrow A$ and noting that $|R^{-1}| = |R|$. $\square$

The reader should review Figure 1.1 to see that all four statements hold for that particular relation.

In our directed graph terminology, $|R|$ is the number of directed arrows in the relation R, and $|R(a)|$ is the number of arrows emanating from a. The average value of $|R(a)|$, taking into account all $a \in A$, is $|R|/|A|$. Hence (1) says that there exists $a \in A$ with at least $|R|/|A|$ arrows emanating from it, and (2) asserts that there exists $a \in A$ with at most $|R|/|A|$ arrows emanating from it. Similarly, (3) states that there exists $b \in B$ with at least $|R|/|B|$ arrows pointing to it, while (4) states that there exists $b \in B$ with at most $|R|/|B|$ arrows pointing to it.

Example. Suppose $A_1, \ldots, A_{50}$ are fifty subsets of a set S with 100 elements, and each A_i contains 40 elements of S. Can we prove the existence of an element of S which is contained in many of the A_i? Define a relation R from $\{A_1, \ldots, A_{50}\}$ to S by letting $(A_i, s) \in R$ if and only if $s \in A_i$. It follows from Theorem 2.1(3) that there exists an $s \in S$ contained in at least $|R|/|S| = 50 \cdot 40/100 = 20$ of the A_i.

In the above application, the relation is not a function. What does the pigeonhole principle say about functions? If $f : A \longrightarrow B$ is a function, then $|f| = |A|$, so $|f|/|A| = 1$; hence (1) and (2) of Theorem 2.1 yield no information. However, (3) and (4) remain important and are restated here in the context of functions.

Theorem 2.2. *If $f : A \longrightarrow B$ is a function, with A and B finite nonempty sets, then the following two statements hold:*

(3) There exists $b \in B$ with $|f^{-1}(b)| \geq |A|/|B|$.

(4) There exists $b \in B$ with $|f^{-1}(b)| \leq |A|/|B|$.

The following generalization of Theorem 2.2 is needed in Chapter 4.

Theorem 2.3. *If $f : A \longrightarrow B$ is a function from a finite nonempty set A to a b-set $B = \{B_1, \ldots, B_b\}$, then the following two statements hold:*

(3) If $|A| = a_1 + \cdots + a_b - b + 1$, then there exists B_i with $|f^{-1}(B_i)| \geq a_i$.

(4) If $|A| = a_1 + \cdots + a_b + b - 1$, then there exists B_i with $|f^{-1}(B_i)| \leq a_i$.

Proof. (3) If there is no such B_i, then $|A| = \sum_{i=1}^{b} |f^{-1}(B_i)| \leq \sum_{i=1}^{b}(a_i - 1) = \sum_{i=1}^{b} a_i - b$, a contradiction.

(4) If there is no such B_i, then $|A| = \sum_{i=1}^{b} |f^{-1}(B_i)| \geq \sum_{i=1}^{b}(a_i + 1) \geq \sum_{i=1}^{b} a_i + b$, a contradiction. $\square$

These theorems are only interesting when $|A| > |B|$, as in the following application.

Example. A popular board game features cards of three suits: cannon, horse, and soldier. A "set" consists of three horses, three soldiers, three cannons, or one card of each suit. It is possible to have four cards without possessing a set, e.g., two horses and two soldiers. Problem: Prove that any five cards contain a set. Solution: Let the three suits be designated by C, H, and S. If the five cards do not include one card of each suit, then at least one suit is absent, say S. Therefore, we may define a function $f : \{a, b, c, d, e\} \longrightarrow \{C, H\}$ from the five cards to their respective suits. (The function isn't necessarily onto.) By Theorem 2.2(3), the preimage of one suit contains at least three cards. These cards constitute a set.

If $|A| = |B| + 1$, then the following special case of the pigeonhole principle results.

Theorem 2.4. *If $f : A \longrightarrow B$ is a function, and $|A| = |B| + 1$, then there exists $b \in B$ with $|f^{-1}(b)| \geq 2$. In other words, $f(a_1) = f(a_2)$ for some distinct $a_1, a_2 \in A$.*

Theorem 2.4 is often paraphrased as follows: "If $k+1$ objects are contained in k pigeonholes, then at least one pigeonhole must contain at least 2 objects." Hence the term *pigeonhole principle*.

Example.   A *lattice point* in the plane is an ordered pair $p = (x, y)$ with integer coordinates x and y. Problem: Suppose five lattice points in the plane p_1, p_2, p_3, p_4, p_5 are given. Prove that the midpoint of the line segment $p_i p_j$ determined by some two distinct lattice points p_i and p_j is also a lattice point. Solution: Define a function $f : \{p_1, p_2, p_3, p_4, p_5\} \longrightarrow \{(0,0), (0,1), (1,0), (1,1)\}$ by mapping p_i to the ordered pair (a_i, b_i), where $a_i \equiv x_i \bmod 2$ and $b_i \equiv y_i \bmod 2$. By Theorem 2.4, some two points p_i and p_j have the same image. To see that p_i and p_j satisfy the requirement of the problem, note that the midpoint of $p_i p_j$ is $((x_i + x_j)/2, (y_i + y_j)/2)$. Both coordinates are integers because x_i and x_j have the same parity and y_i and y_j have the same parity.

The value 5 in the problem above is "best possible" in the sense that one can find four lattice points determining no lattice midpoint, e.g., $(0,0), (0,1), (1,0), (1,1)$. Is 20 in the first example the best possible result?

Example.   The *centroid* of three points $p_i = (x_i, y_i)$, $p_j = (x_j, y_j)$, $p_k = (x_k, y_k)$ is $((x_i + x_j + x_k)/3, (y_i + y_j + y_k)/3)$. What is the minimum number n of lattice points, some three of which must determine a lattice point centroid? We show that $n \le 13$ by defining a function $f : \{p_1, \ldots, p_{13}\} \longrightarrow \{0, 1, 2\}$ which maps p_i to the residue class modulo 3 of its first coordinate. By Theorem 2.2(3), some five lattice points, say p_1, p_2, p_3, p_4, p_5, have the same image. By the analysis of the suits in the board game mentioned above, some three of these points, say p_1, p_2, p_3, have second coordinate residues 000, 111, 222, or 012. These three lattice points determine a lattice point centroid. Therefore n exists and satisfies $n \le 13$.

The determination of the minimum n which forces the existence of three points determining a lattice point centroid is a more difficult matter. It turns out that the minimum value is $n = 9$. For simplicity, the entire argument is carried out modulo 3. Thus there are just nine possible ordered pairs from which to choose. The following list of ordered pairs (modulo 3) shows that $n > 8$:

$$(0,0), (0,0), (1,0), (1,0), (0,1), (0,1), (1,1), (1,1).$$

In order to prove $n \leq 9$, we must show that any nine points include three whose first coordinates and second coordinates are of the form 000, 111, 222, or 012. The nine possible ordered pairs are conveniently represented by the nine non-ideal points of the order 3 projective plane of Figure 10.1. The twelve lines of the figure (excluding the line at infinity) correspond to triples of points which determine a lattice point centroid. If any ordered pair is chosen three times, these three ordered pairs determine a lattice point centroid. Therefore, let us assume that each ordered pair is chosen at most twice, and hence at least five different ordered pairs are chosen. By shuffling the rows and columns of the figure (if necessary), we may assume that three of the points are $(0,0)$, $(1,0)$, and $(1,1)$. If no three points are collinear, then we may not choose the points $(2,0)$, $(0,2)$, or $(2,2)$. This means that we must choose two of the three points $(1,1)$, $(2,1)$, and $(1,2)$. But any of these choices gives three collinear points: $(0,1)$, $(1,1)$, and $(2,1)$; $(1,0)$, $(1,1)$, and $(2,1)$; or $(0,0)$, $(1,2)$, and $(2,1)$.

Open problem 1.   *What is the minimum number n of lattice points in the plane, some four of which must determine a lattice point centroid?*

The four points must have x and y coordinates of the form 0000, 1111, 2222, 3333, 0022, 0013, 0112, 0233, 1133, or 1223. A simple application of the pigeonhole principle shows that of 49 points, some four have both coordinates of type 0000, 1111, 2222, or 3333. Thus the minimum number n called for in the problem satisfies $n \leq 49$. The set $\{(0,0),(1,0),(0,1),(1,1)\}$, taken with multiplicity three, shows that $n > 12$.

In general, define $n(k)$ to be the smallest value of n such that, given any n lattice points in the plane, some k have a lattice point centroid. The lower bound given above for the $k = 4$ case generalizes to show $n > 4(k-1)$.

Open Problem 2.   *Find a formula for n(k) for all k.*

The d-dimensional generalization of the above problem calls for the minimum n such that, given any n lattice points in $\mathcal{R}^d$ (ordered d-tuples of integers), some k determine a lattice point centroid. Let $n(k, d)$ be the minimum such n. The existence of $n(k, d)$ is guaranteed by the pigeonhole principle, and, in fact, the pigeonhole principle yields an upper bound for $n(k, d)$ on the order of k^{d+1}. The set of d-tuples each of whose coordinates is 0 or 1, taken with multiplicity $k - 1$, establishes the lower bound $n(k, d) > 2^d(k - 1)$.

Open Problem 3. *Find a formula for $n(k, d)$ for all k, d.*

This question is known as the *Gitterpunktproblem* ("lattice point problem"). Not much is known about the function $n(k, d)$. Trivially, $n(1, d) = 1$. In 1949 P. Erdös, A. Ginzburg, and A. Ziv showed $n(k, 1) = 2k - 1$. It has since been shown that $n(3, 3) = 19$. See Problem 6298, *American Mathematical Monthly* **89** (1982) 279-280. As indicated above, no formula is known for $n(k, 2)$, although we have shown that $n(2, 2) = 5$ and $n(3, 2) = 9$. Exercise 2.4 calls for a proof that $n(2, d) = 2^d + 1$.

We now give a pigeonhole principle proof of a very old but interesting result in number theory. For alternative proofs the reader is referred to Niven, Zuckerman, and Montgomery (1991).

Theorem 2.5. *For all real numbers α and $n \in \mathcal{N}$, there exist $p, q \in \mathcal{N}$ with $1 \le q \le n$ and $|\alpha - p/q| < 1/qn \le 1/q^2$.*

Proof. Define

$$f : \mathcal{N}_{n+1} \longrightarrow \left\{ \left[0, \frac{1}{n}\right), \left[\frac{1}{n}, \frac{2}{n}\right), \ldots, \left[\frac{n-1}{n}, 1\right) \right\}$$

by letting $f(j)$ be the subinterval of $[0, 1)$ which contains $j\alpha - \lfloor j\alpha \rfloor$. ($\lfloor x \rfloor$ is the greatest integer less than or equal to x.) The pigeonhole principle (Theorem 2.4) implies the existence of j, k ($j > k$) with $f(j) = f(k)$. According to the definition of the function f, there exists a positive integer p with $|j\alpha - k\alpha - p| < 1/n$. Letting $q = j - k$ (so that $1 \le q \le n$), the inequality becomes $|\alpha - p/q| < 1/qn \le 1/q^2$. $\square$

Theorem 2.5 is used to ensure good rational approximations to irrational numbers α. For example, taking $\alpha = \pi$ and $n = 10$, the theorem guarantees the existence of a rational number p/q with $|\pi - p/q| < 1/10q \le 1/q^2$. In fact, the well-known approximation $\frac{22}{7}$ satisfies the inequality. Can the reader find an approximation to $\alpha = \sqrt{2}$ with $n = 10$?

In the preceding versions of the pigeonhole principle we have assumed that both the domain and the codomain are finite sets. If the domain is infinite, then the following highly useful infinitary pigeonhole principle results.

Theorem 2.6. *(Infinitary pigeonhole principle). If $f : A \longrightarrow B$ is a function from a countably infinite set A to a finite nonempty set B, then there exists $b \in B$ with $f^{-1}(b)$ countably infinite.*

The proof is left as an exercise. An application of the infinitary pigeonhole principle is given in the proof of Theorem 3.2.

2.2. GRAPH THEORY

Graph theory began in 1736 when L. Euler (1707–1783) solved the famous Königsberg bridge problem concerning a certain system of seven bridges over the river Pregel. See Hartsfield and Ringel (1994). In the last thirty years there has been an explosion in graph theory research and applications. Today, there are many areas of graph theory, research including algebraic graph theory, extremal graph theory, and topological graph theory. Within combinatorics, graph theory is closely related to design theory, Ramsey theory, and coding theory. In this section we give some basic definitions and an indication of the deeper results of graph theory which will be studied in the later chapters.

A *graph G* is an ordered pair (V, E), consisting of a *vertex set V* and an *edge set* $E \subseteq [V]^2$. Vertices are also called *points* or *nodes*. Edges are also called *lines* or *arcs*. In our definition of graph there are no loops or multiple edges. In a drawing of a graph, two vertices x and y are joined by a line if and only if $\{x, y\} \in E$. Two vertices joined by a line are said to be *adjacent*; if they are not joined by a line, they are *nonadjacent*. If $|V| = p$ and $|E| = q$, then we say that G has *order p* and *size q*.

Certain graphs occur so frequently that they require names. The *complete graph* K_n consists of n vertices and all $\binom{n}{2}$ possible edges. The *complete bipartite graph* $K_{m,n}$ consists of a set A of m points, a set B of n points, and all the mn edges between A and B. The *complete infinite graph* K_∞ contains a countably infinite set of points and all possible edges. Likewise, the *complete infinite bipartite graph* $K_{\infty,\infty}$ contains countably infinite sets A and B and all edges between A and B. The *cycle* C_n consists of n vertices connected by n edges in a cyclical fashion. The *path* P_n is C_n minus an edge. Figure 2.1 illustrates some of these graphs. For a general reference on graph theory, the reader is refered to Bollobás (1979).

The simplest theorem of graph theory is called the handshake theorem. Suppose a finite number of people meet and some shake hands. Assume that no person shakes his or her own hand and furthermore no two people shake hands more than once. Each person tells how many times he or she shook hands. The handshake theorem says that the sum of these numbers is an even number. In graph theory terms, the people are represented by a vertex set V, and two vertices are adjacent if and only if the corresponding people shake hands. The number of vertices to which a given vertex v is adjacent is the *degree* $\delta(v)$ of v. (This is the number of people who shake hands with person v.) Consider the sum $\sum_{v \in V} \delta(v)$. Because each edge contributes exactly 2 to this sum, $\sum_{v \in V} \delta(v) = 2q$, an even number.

Theorem 2.7. *(Handshake theorem). In any graph G with a finite number of vertices, $\sum_{v \in V} \delta(v)$ is an even number.*

Corollary 2.8. *In any graph G with a finite number of vertices, the number of vertices with odd degree is an even number.*

Here is a similar theorem whose proof uses the pigeonhole principle.

Theorem 2.9. *In any graph G with a finite number of vertices, some two vertices have the same degree.*

Proof. Suppose G has p vertices. Then each vertex has degree equal to one of the numbers $0, \ldots, p - 1$. However, it is impossible for G to have both a vertex of degree 0 and a vertex of degree $p - 1$. Therefore the list of degrees of the p vertices contains at most $p - 1$ different numbers. By the pigeonhole principle, some two vertices have the same degree. $\square$

There is a certain situation in which the sum $\sum_{v \in V} \delta(v)$ is particularly easy to compute, namely, when $\delta(v)$ is the same for all vertices. In this case we say that G is *regular* of *degree* $\delta(v)$. Note that complete graphs and cycles are regular.

If G is any finite graph, the *independence number* $\alpha(G)$ is the maximum possible number of pairwise nonadjacent vertices of G. The *chromatic number* $\chi(G)$ of G is the minimum number of colors in a coloring of the vertices of G with the property that no two adjacent vertices share the same color.

Here is another simple theorem proved by the pigeonhole principle.

Theorem 2.10. *In any graph G with p vertices, $p \leq \alpha(G)\chi(G)$.*

Proof. Consider the vertices of G as partitioned into $\chi(G)$ color classes. By the pigeonhole principle, one of the classes must contain at least $p/\chi(G)$ vertices, and these vertices are pairwise nonadjacent. Thus $\alpha(G) \geq p/\chi(G)$ and the result follows immediately. $\square$

Equality in the above theorem holds, for example, when G consists of the vertices and edges of a cube.

The famous "four color theorem," proved in 1976 by K. Appel and W. Haken, is the statement that $\chi(G) \leq 4$ for any planar graph G. Combining this result with Theorem 2.10 we arrive at the relation $\alpha(G) \geq p/4$ for any planar graph G. We can turn a planar graph into a planar map by placing a territory at each vertex and allowing two territories to share a common boundary when the two vertices in the graph are adjacent. In terms of maps, Appel and Haken's result is that every map can be colored with four colors so that no two bordering territories have the same color. It follows from Theorem 2.10 that any planar map on p vertices contains at least $p/4$ territories no two of which share a border.

A *path* in a graph G from vertex v_0 to vertex v_n is a sequence of distinct edges

$$\{v_0, v_1\}, \{v_1, v_2\}, \ldots, \{v_{n-1}, v_n\}.$$

The path is *simple* if the vertices $v_1, v_2, \ldots, v_n$ are distinct. A *circuit* is a path from v to v for some vertex v. A *simple circuit* is a cycle. We say that G is *connected* if there is a path between every two vertices. Note that each of the graphs in Figure 2.1 is connected.

We say that two graphs G_1 and G_2 are *isomorphic* if there is a bijection from V_1 to V_2 which preserves adjacency. Note that in Figure 2.1, C_4 is isomorphic to $K_{2,2}$. We normally regard two isomorphic graphs as the same graph. In Chapter 8 we develop a formula for the number of nonisomorphic graphs with a given number of vertices.

2.3. EXTREMAL GRAPHS

We round off the second chapter by presenting two applications of the pigeonhole principle, which, although simple, introduce the vital ideas of Ramsey theory. The first, a special case of Turán's theorem called Mantel's theorem, presages Ramsey's theorem (Section 4.1), while the second, concerning partitions of the plane, foreshadows Van der Waerden's theorem (Section 4.6).

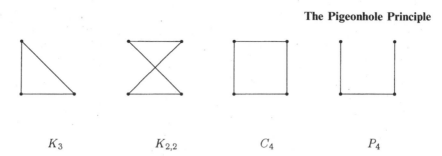

K_3 $K_{2,2}$ C_4 P_4

Figure 2.1. A complete graph, a complete bipartite graph, a cycle, and a path.

Here we discuss an *extremal property* of graphs. How many edges are possible in a triangle-free graph G on $2n$ vertices? Certainly, G can have n^2 edges without containing a triangle: just let G be the complete bipartite graph $K_{n,n}$, consisting of two sets of n vertices each and all the edges between the two sets. Indeed, n^2 turns out to be the maximum possible number of edges. That is, if G contains $n^2 + 1$ edges, then G contains a triangle. This we prove by induction, using the pigeonhole principle.

Theorem 2.11. *(Mantel, 1907). If a graph G of order 2n contains $n^2 + 1$ edges, then G contains a triangle.*

Proof. If $n = 1$, then G can't have $n^2 + 1$ edges; hence the statement is vacuously true. Assuming the result for n, we now consider a graph G on $2(n + 1)$ vertices with $(n + 1)^2 + 1$ edges. Let x and y be adjacent vertices in G, and let H be the restriction of G to the other $2n$ vertices. If H contains more than n^2 edges, then we are finished by the induction hypothesis. Suppose H has at most n^2 edges, and therefore at least $2n + 1$ edges emanate from x and y to vertices in H. By the pigeonhole principle (Theorem 2.2), there exists an edge from x and an edge from y to the same vertex z in H. Hence G contains the triangle xyz. $\square$

We give a second proof that does not use the pigeonhole principle.

Proof. Assume that G has no triangles. We assign a nonnegative weight w_i to each vertex of G so that $\sum w_i = 1$. Let $M = \sum w_i w_j$, where the sum is taken over all edges ij. We seek to maximize M. One way of assigning the weights is to let $w_i = 1/(2n)$ for each i. This gives $M = q/(4n^2)$. Now suppose that i and j are two nonadjacent vertices, W_i and W_j are the total weights of vertices connected to i and j, respectively, and $W_i \geq W_j$. Notice that if we shift all of the weight from j to i we do not decrease the value of $\sum w_i w_j$, for $w_i W_i + w_j W_j \leq (w_i + w_j) W_i$. It follows that M is maximized when all of the weight is concentrated on two adjacent vertices, say i and j. Using calculus, it can be shown that $w_i w_j$, subject to $w_i + w_j = 1$, attains a maximum value of $\frac{1}{4}$ (when $w_i = w_j = \frac{1}{2}$). Thus $\frac{1}{4} \geq q/(4n^2)$, from which the result follows immediately. $\square$

2.4. COLORINGS OF THE PLANE

Suppose the plane is partitioned into two (disjoint) subsets G and R (green and red). We will show that one of the two subsets contains the vertices of a Euclidean rectangle with sides parallel to the coordinate axes. In fact, partitioning the whole plane is unnecessary. The same result follows if we partition just the 21 lattice points of $\mathcal{N}_7 \times \mathcal{N}_3$ into two subsets, so let us assume only that. We say that each lattice point is "colored" either G or R.

Theorem 2.12. *If the 21 lattice points of $\mathcal{N}_7 \times \mathcal{N}_3$ are colored G and R, there exist four points, all the same color, lying on the vertices of a rectangle with sides parallel to the coordinate axes.*

Proof. Each column of three points in this lattice contains three points of color G, two G's and one R, two R's and one G, or three R's. For the moment, the relevant fact is that there is a majority of G's or a majority of R's in each column. Let us refer to a column as a G-majority or an R- majority column. By the pigeonhole principle (Theorem 2.2), some four columns are G-majority columns or some four columns are R-majority columns. Without loss of generality, suppose there are four G-majority columns. We will show that there are four points all colored G which are the vertices of a rectangle. If any of the four G-majority columns contain three points colored G, then we handicap ourselves by changing the color of an arbitrary point to R. (Our result will follow even with this handicap.) Now we have four columns which each contain exactly two points colored G and one point colored R. There are three possible patterns for the configuration of the points: GGR, GRG, and RGG. By the pigeonhole principle (Theorem 2.4), there are two columns with the same color pattern. The four G's in these columns are the vertices of a rectangle with sides parallel to the coordinate axes. $\square$

It is easy to see that neither the lattice $\mathcal{N}_7 \times \mathcal{N}_2$ nor $\mathcal{N}_6 \times \mathcal{N}_3$ is sufficient to force the existence of four monochromatic points on the vertices of a rectangle with sides parallel to the coordinate axes. See

Exercise 2.21. Thus the lattice $\mathcal{N}_7 \times \mathcal{N}_3$ is *minimal* with respect to this property.

The corresponding minimality problem with 'rectangle' replaced by 'square' is much more difficult, as the following open problem attests. See Exercise 4.24.

Open Problem 4. *Find the minimum n such that if the n^2 lattice points of $\mathcal{N}_n \times \mathcal{N}_n$ are two-colored, there exist four points of one color lying on the vertices of a square with sides parallel to the coordinate axes.*

Notes

P. G. L. Dirichlet (1805–1859) was the first mathematician to explicitly use the pigeonhole principle in proofs. He referred to it as the drawer principle.

With reference to Theorem 2.5, suppose q is an integer satisfying the inequality $|\alpha - p/q| < 1/qn \le 1/q^2$ for a given n. If α is irrational, then $|\alpha - p/q|$ is not 0, so that the inequality is violated for n sufficiently large. Thus a new value of n gives rise to a new value of q, and so on. Therefore, if α is irrational, there are infinitely many pairs p, q for which $|\alpha - p/q| \le 1/q^2$. A theorem of A. Hurwitz states that this remains true when the quantity on the right side of the inequality is decreased by a factor of $\sqrt{5}$, but that $\sqrt{5}$ is best possible. Furthermore, if α is irrational and $x > 2$, then there are only finitely many pairs p, q for which $|\alpha - p/q| \le 1/q^x$. See Niven, Zuckerman, and Montgomery (1991).

The word "graph" was first used in mathematics in an 1877 paper by J. Sylvester. In 1936 D. König wrote the first book on graph theory, *Theorie der endlichen und unendlichen Graphen.*

The special case of Turán's theorem (1941) was proved by W. Mantel in 1907.

Exercises

2.1 Using the pigeonhole principle, show that some positive integral power of 17 ends in 0001 (base ten). Prove the same proposition using Euler's theorem.

2.2 (Putnam Competition, 1978) Let A be any set of 20 distinct integers chosen from the arithmetic progression

$$1, 4, 7, \ldots, 100.$$

Prove that there must be two distinct integers in A whose sum is 104.

[Hint: Let the pigeonholes be the sets $\{1\}, \{4, 100\}, \ldots, \{49, 55\}, \{52\}$.]

2.3 (Putnam Competition, 1971) Let there be given nine lattice points (points with integer coordinates) in three-dimensional Euclidean space. Show that there is a lattice point on the interior of one of the line segments joining two of these points.

2.4 Prove that $n(2, d) = 2^d + 1$. Also show that $n(3, 3) > 18$.

2.5 Suppose $A_1, \ldots, A_{100}$ are subsets of a finite set S, each with $|A_i| > \frac{2}{3}|S|$. Prove that there exists $x \in S$ with x contained in at least 67 of the A_i. Show that 67 is the best possible result. [Hint: To show that this result is best possible, let $S = \{1, \ldots, 100\}$ and take A_1 to be any subset of S with 67 elements. Let the other A_i be cyclic shifts of A_1. In this system of sets, each $x \in S$ is contained in exactly 67 of the A_i.]

2.6 Show that if S is a subset of $\{1, \ldots, 2n\}$ and $|S| > n$, then there exist $x, y \in S$ with x and y relatively prime.

2.7 Show that if S is a subset of $\{1, \ldots, 2n\}$ and $|S| > n$, then there exist $x, y \in S$ with x a divisor of y.

2.8 (Putnam Competition, 1964) Let S be a set of $n > 0$ elements, and let $A_1, A_2, \ldots, A_k$ be a family of distinct subsets, with the property that any two of these subsets meet. Assume that no other subset of S meets all of the A_i. Prove that $k = 2^{n-1}$.

[Note: This problem is generalized in many interesting ways in Anderson (1987).]

2.9 Prove Theorem 2.6 (infinitary pigeonhole principle). [Hint: Assume $f^{-1}(b)$ is finite for all $b \in B$, and obtain a contradiction.]

2.10 Show that Theorem 2.9 does not hold for infinite graphs.

2.11 Find two nonisomorphic graphs with $p = 12$, $\alpha(G) = 3$, and $\chi(G) = 4$. This shows that the upper bound of Theorem 2.10 may be met by nonisomorphic graphs.

2.12 Use the infinitary pigeonhole principle to prove that if G is a countably infinite graph, then at least one of $\alpha(G)$ and $\chi(G)$ must be infinite.

2.13 (L. Euler, 1736) An *Euler circuit* is a circuit containing all edges of G. Show that a connected graph G has an Euler circuit if and only if each vertex of G has even degree.

2.14 (L. Euler, 1752) A graph is *planar* if it can be drawn in the plane with lines intersecting only at the vertices. A planar graph partitions the plane into regions called *faces*. Let F be the number of faces, E the number of edges, and V the number of vertices of a planar graph. Show that $E + 2 = V + F$.

2.15 Show that K_5 is not planar. Show that $K_{3,3}$ is not planar.

2.16 Show that if G is a graph with $\delta(v) \geq p/2$ for every vertex v, then G is connected.

2.17 (G. A. Dirac, 1952) Show that under the hypothesis of the previous exercise, G contains a subgraph isomorphic to C_p. Such a subgraph is called a *Hamiltonian circuit*, after the mathematician W. R. Hamilton.

2.18 A *tree* is a connected graph with no cycles. Prove that in a tree with p vertices and q edges, $p = q + 1$. Deduce that if G is a graph with $p < q + 1$, then G contains cycles, and if $p > q + 1$, then G is disconnected.

2.19 Show that, up to isomorphism, $K_{n,n}$ is the only triangle-free graph with $2n$ vertices and n^2 edges.

2.20 Exhibit colorings which show that neither the lattice $\mathcal{N}_7 \times \mathcal{N}_2$ nor $\mathcal{N}_6 \times \mathcal{N}_3$ is sufficient to force the existence of four monochromatic points on the vertices of a rectangle with sides parallel to the coordinate axes.

2.21 Let L_1 be a two-row array of positive integers

$$\begin{array}{l} a_1 \quad a_2 \dots a_m \\ b_1 \quad b_2 \dots b_m \end{array},$$

where the a_i are distinct integers written in increasing order. Let $c_1, \dots, c_n$ be the list of all integers which occur in $L_1 (c_1 \leq c_2 \leq \dots \leq c_n)$, and, for each i, $1 \leq i \leq m$, let d_i be the number of occurrences of c_i in L_1. Let L_2 be the array

$$\begin{array}{l} c_1 \quad c_2 \dots c_n \\ d_1 \quad d_2 \dots d_n \end{array}.$$

For example, if L_1 is the array

$$\begin{array}{ccccccc} 1 & 2 & 5 & 6 & 8 & 10 & 11 \\ 3 & 3 & 1 & 4 & 1 & 3 & 6 \end{array},$$

then L_2 is the array

$$
\begin{array}{ccccccccc}
1 & 2 & 3 & 4 & 5 & 6 & 8 & 10 & 11 \\
3 & 1 & 3 & 1 & 1 & 2 & 1 & 1 & 1
\end{array}
$$

Starting with any array L_1, L_2 is created as indicated above. Then the operation is repeated on L_2 to form a new array L_3, and so on. Show that the number of distinct arrays produced in this matter is always finite. We say that each sequence of arrays eventually "goes into a loop." Show that a loop always consists of one, two, or three arrays.

3

Sequences and Partial Orders

The pigeonhole principle of Chapter 2 guarantees that not every element in a set is below average and not every element is above average. In Chapter 3 we apply the principle to two types of mathematical structures: sequences and partial orders. Our goal is to show that arbitrary sequences and partial orders contain highly nonrandom substructures. This is an example of a basic result of existential combinatorics: complete disorder is impossible.

3.1. THE ERDÖS–SZEKERES THEOREM

We now address the proposition cited at the beginning of Part I. A *finite sequence* $S = \{a_1, \ldots, a_n\}$ of real numbers is a function $a : \{1, \ldots, n\} \longrightarrow \mathcal{R}$, for some $n \in \mathcal{N}$, defined by $a(k) = a_k$ for all $k \in \mathcal{N}_n$. For example, $\{1, 0, \sqrt{2}\}$ is a finite sequence of real numbers with three terms. An *infinite sequence* $S = \{a_1, a_2, \ldots\}$ of real numbers is a function $a : \mathcal{N} \longrightarrow \mathcal{R}$ defined by $a(k) = a_k$ for all $k \in \mathcal{N}$. For example, the function $a(k) = k^2$ defines the infinite sequence $S = \{1, 4, 9, \ldots\}$. We often drop the braces when we write a sequence.

A sequence (finite or infinite) is *increasing* (or *strictly increasing*) if $a_i < a_j$ for all $i < j$; *decreasing* (or *strictly decreasing*) if $a_i > a_j$ for all $i < j$; *monotonically increasing* if $a_i \le a_j$ for all $i < j$; *monotonically decreasing* if $a_i \ge a_j$ for all $i < j$; and *monotonic* if it is either monotonically increasing or monotonically decreasing. Thus the sequence $\{1, 1, 0, -1\}$ is monotonically decreasing and the sequence $\{1, 2, 3, 4, \ldots\}$ is strictly increasing.

A sequence $\{b_1, \ldots, b_m\}$ is a *subsequence* of $\{a_1, \ldots, a_n\}$ if there exists a one-to-one function $f : \{1, \ldots, m\} \longrightarrow \{1, \ldots, n\}$ for which $f(i) < f(j)$ whenever $i < j$, and $b(i) = a(f(i))$ for all $i \in \mathcal{N}_m$. For example, $\{1, 2, 3, 2\}$ is a subsequence of $\{1, 4, 2, 3, 5, 2\}$.

Theorem 3.1. *(The Erdös–Szekeres theorem). Let* $m, n \in \mathcal{N}$. *Any sequence* $S = \{a_1, \ldots, a_{mn+1}\}$ *of* $mn + 1$ *real numbers contains a monotonically increasing subsequence of* $m + 1$ *terms or a monotonically decreasing subsequence of* $n + 1$ *terms (or both).*

Proof. Assume that S does not contain a subsequence of either type. We define a function $f : \mathcal{N}_{mn+1} \longrightarrow \mathcal{N}_m \times \mathcal{N}_n$ by setting $f(t) = (x_t, y_t)$, where x_t is the greatest number of terms in a monotonically increasing subsequence of S beginning with a_t, and y_t is the greatest number of terms in a monotonically decreasing subsequence beginning with a_t. By the pigeonhole principle (Theorem 2.4), there exist $j, k \in \{1, \ldots, mn + 1\}$, $j < k$, with $f(j) = f(k)$. It follows that if $a_j \leq a_k$, then $x_j > x_k$, while if $a_j \geq a_k$, then $y_j > y_k$. Both inequalities contradict $f(j) = f(k)$. Therefore, our original assumption is false, and S *does* contain a monotonically increasing subsequence of $m + 1$ terms or a monotonically decreasing subsequence of $n + 1$ terms. $\square$

Example. Taking $m = 3$ and $n = 3$, the Erdös–Szekeres theorem guarantees that a sequence S of $mn + 1 = 10$ real numbers contains a monotonic subsequence of 4 terms. If the ten terms of S are distinct, then of course the subsequence will be strictly increasing or strictly decreasing. Thus, in our example at the beginning of Part I, the sequence $5, 8, -1, 0, 2, -4, 1, 7, 6$ contains the strictly increasing subsequence $-1, 0, 2, 7$.

The expression $mn + 1$ in Theorem 3.1 is best possible in the sense that there exists a sequence of mn real numbers which contains neither a monotonically increasing subsequence of $m + 1$ terms or a monotonic decreasing subsequence of $n + 1$ terms. We form such a sequence by concatenating n sequences of m increasing terms in the following manner. For each $j \in \mathcal{N}_n$, let $S_j = \{a_{1j}, \ldots, a_{mj}\}$ be a monotonically increasing sequence of m real numbers, with $a_{hj} > a_{lk}$ whenever $j < k$. The reader should verify that the sequence

$$S = \{a_{11}, \ldots, a_{m1}, a_{12}, \ldots, a_{m2}, \ldots, a_{1n}, \ldots, a_{mn}\}$$

contains no monotonic subsequence of the desired length. The question of the number of such sequences is answered by the theory of Young tableaux. See Notes and Chapter 7.

The set of real numbers $\mathcal{R}$ is *linearly ordered*. That is, for any two distinct real numbers x and y, either $x < y$ or $y < x$. Dilworth's lemma of the next section generalizes the Erdös–Szekeres theorem to partially ordered sets, i.e., sets in which two elements may or may not be comparable.

We conclude this section with an infinitary version of the Erdös–Szekeres theorem. The proof is a nice double application of the infinitary pigeonhole principle.

Theorem 3.2. *(The Erdös–Szekeres theorem, infinitary version) Any infinite sequence $S = \{a_1, a_2, \ldots\}$ of real numbers contains an infinite monotonic subsequence.*

Proof.   We describe a subsequence $T = \{b_1, b_2, \ldots\}$ of S inductively as follows. Let $b_1 = a_1$. By the infinitary pigeonhole principle (Theorem 2.6), there is an infinite subset $S^2 = \{a_2^2, a_3^2, a_4^2, \ldots\}$ of $S - \{a_1\}$ all of whose elements are greater than or equal to b_1, or all of whose elements are less than or equal to b_1. Let $b_2 = a_2^2$. Continuing in this manner, we find an infinite subset $S^3 = \{a_3^3, a_4^3, a_5^3, \ldots\}$ of $S^2 - \{a_2^2\}$, all of whose elements are greater than or equal to b_2, or all of whose elements are less than or equal to b_2. This process defines the sequence T. Each $b_i \in T$ is either greater than or equal to all elements following it, or less than or equal to all elements following it. Again, by the infinitary pigeonhole principle, there is a subsequence $T' \subseteq T$, each of whose elements is greater than or equal to those following it, or each of whose elements is less than or equal to those following it. Therefore, T' is an infinite monotonic subsequence of S. $\square$

3.2. DILWORTH'S LEMMA

A partial order $\leq$ on X is a *total order* (or *linear order*) if every two elements of X are comparable. For example, the usual less than or equal to relation is a linear order on the set $\mathcal{N}$. If $\leq$ is a partial order on X, and Y is a subset of X in which every two elements are comparable, then Y is a *chain* (of X). If no two distinct elements of Y are comparable, then Y is an *antichain* (of X). The *length* of $\leq$ is the greatest number of elements in a chain of X, and the *width* of $\leq$ is the greatest number of elements in an antichain. For example, Figure 3.1 is the directed graph representation of the partial order

$$\leq = \{(3,1),(4,1),(4,3),(3,2),(4,2),(4,5),(6,5),(7,5),(7,6)\}$$

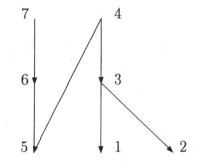

Figure 3.1. A partial order of width 3 and length 3.

on the set $X = \{1, 2, 3, 4, 5, 6, 7\}$. The arrows required for reflexivity and transitivity are suppressed in the diagram. The length and width of $\leq$ are both 3; for instance, $\{1, 3, 4\}$ is a chain of length 3, and $\{1, 2, 6\}$ is an antichain of size 3.

The length and width of a partial order are related to the size of the underlying set by the following result of R. P. Dilworth.

Theorem 3.3. *(Dilworth's lemma). In any partial order on a set X of $mn + 1$ elements, there exists a chain of length $m + 1$ or an antichain of size $n + 1$.*

Proof. Suppose there is no chain of size $m + 1$. Then we may define a function $f : X \longrightarrow \{1, \ldots, m\}$ with $f(x)$ equal to the greatest number of elements in a chain with greatest element x. By the pigeonhole principle (Theorem 2.2(3)), some $n + 1$ elements of X have the same image under f. By the definition of f, these elements are incomparable; that is, they form an antichain of size $n + 1$. $\square$

As an example of Dilworth's lemma, we consider again the partial order of Figure 3.1. The size of X is $|X| = 7 = 2 \cdot 3 + 1$. Therefore Dilworth's lemma guarantees a chain of $2 + 1$ elements or an antichain of $3 + 1$ elements; and we have remarked that there is, indeed, a chain of length 3. Notice the similarity between Dilworth's lemma and the Erdös–Szekeres theorem (Theorem 3.1). In each case, the hypothesis concerns a set of $mn + 1$ elements and the conclusion concerns subsets of sizes $m + 1$ and $n + 1$. These similarities are no coincidence. In fact, the Erdös–Szekeres theorem may be proved as a corollary of Dilworth's lemma. Let $S = \{a_1, \ldots, a_{mn+1}\}$ be a sequence of $mn + 1$ real numbers. Define a partial order $\leq_S$ on S by setting $a_i \leq_S a_j$ if $a_i \leq a_j$ and $i \leq j$. Dilworth's lemma guarantees a chain of $m + 1$ elements (corresponding to a monotonically increasing subsequence of $m + 1$ terms) or an antichain of $n + 1$ elements (corresponding to a monotonically decreasing subsequence of $n + 1$ terms).

Just as the Erdös–Szekeres theorem is a best possible result, so also is Dilworth's lemma. In the exercises, the reader is asked to furnish an example of a partial order on $|X| = mn$ elements with length m and width n.

Dilworth's lemma is sometimes easier to apply in the following equivalent form.

Theorem 3.4.   *Let $|X| = n$, and let $\leq$ be a partial order on X with length l and width w. Then $n \leq lw$.*

Proof.   Suppose, to the contrary, that $n \geq lw + 1$. Then, by Dilworth's lemma, there is a chain of length $l + 1$ or an antichain of size $w + 1$; these results contradict the definition of l or w. $\square$

The above proof shows that Theorem 3.4 follows from Dilworth's lemma. In the exercises, the reader is asked to show how Dilworth's lemma follows from Theorem 3.4.

Let us use the inequality in Theorem 3.4 to investigate the collection $X(t)$ of subsets of the t-element set $\mathcal{N}_t$, ordered by set inclusion. Clearly, $n = 2^t$. Furthermore, $l = t + 1$, because any chain of greatest length starts with the empty set and adds one element at a time until $\mathcal{N}_t$ is exhausted. What is w? This question is answered by Sperner's theorem of the next section, but for now we note that Dilworth's inequality yields $2^t \leq (t + 1)w$. This inequality is weak. Indeed, the binomial coefficient $\binom{t}{\lfloor t/2 \rfloor}$ satisfies $\binom{t}{\lfloor t/2 \rfloor} \geq 2^t/(t + 1)$, and the subsets of size $\lfloor t/2 \rfloor$ form an antichain establishing $w \geq 2^t/(t + 1)$.

The name Dilworth's lemma suggests that there might be a Dilworth's theorem, which is the case.

Theorem 3.5.   *(Dilworth's theorem). If $\leq$ is a partial order on a set X, with length l and width w, then X can be partitioned into l antichains or w chains.*

We prove that X can be partitioned into l antichains. Define $f : X \longrightarrow \{1, \ldots, l\}$ by letting $f(x)$ be the maximum number of elements in a chain with greatest element x. The preimage of each $y \in \{1, \ldots, l\}$ is an antichain. For the proof that X can be partitioned into w chains, we refer the reader to Bollobás (1979). Considering Figure 3.1 again, we find that X may be partitioned into three chains: $\{1,3,4\}$, $\{2\}$, $\{5,6,7\}$, and into three antichains: $\{1,2,5\}$, $\{3,6\}$, $\{4,7\}$.

As the reader probably suspects, there is an infinitary version of Dilworth's lemma. The proof is an exercise.

Theorem 3.6. *(Dilworth's lemma, infinitary version). A partial order on $\mathcal{N}$ yields an infinite chain or an infinite antichain.*

3.3. SPERNER'S THEOREM

Let $X(t)$ be the collection of subsets of the t-element set $\mathcal{N}_t$, and let $\subseteq$ be the containment partial order on $X(t)$. For instance, if $t = 3$, then $X(t)$ consists of eight elements: $\emptyset$, $\{1\}$, $\{2\}$, $\{3\}$, $\{1,2\}$, $\{1,3\}$, $\{2,3\}$, and $\{1,2,3\}$. Some examples of containment are $\{1,2\} \subseteq \{1,2,3\}$, $\emptyset \subseteq \{3\}$, and $\{1,3\} \subseteq \{1,3\}$. The cardinality of $X(t)$ is 2^t. What are the length and width of $\subseteq$? As we remarked in the previous section, the length is $l = t + 1$, because the longest chains start with $\emptyset$ and include one new element at each step until all t elements are included. The width w of $\subseteq$ is the subject of Sperner's theorem. Remember that an antichain of $X(t)$ is a collection of subsets of $\mathcal{N}_t$ none of which contains another.

E. Sperner proved the following theorem in 1928. We give a simpler proof essentially due to D. Lubell. See Gessel and Rota (1987). For a proof of Sperner's theorem using the probabilistic method, the reader may consult Alon and Spencer (1992). Recall that the binomial coefficient $\binom{n}{k} = n!/[k!(n-k)!]$ is the number of k-subsets of an n-set.

Theorem 3.7. *(Sperner's theorem). The cardinality m of an antichain A of subsets of $\mathcal{N}_t$ satisfies $m \leq \binom{t}{\lfloor t/2 \rfloor}$. Therefore, as the $\binom{t}{\lfloor t/2 \rfloor}$ subsets of size $\lfloor t/2 \rfloor$ form an antichain, the width of $\subseteq$ is $w = \binom{t}{\lfloor t/2 \rfloor}$.*

Proof. Let $A = \{A_1, \ldots, A_m\}$, and let $|A_i| = \alpha_i$ for each $i \in \mathcal{N}_m$. For each A_i there are $\alpha_i!(t - \alpha_i)!$ chains of length $t + 1$ which contain A_i. (Here we are using *enumeration* techniques: such chains commence with the empty set, add one element at a time until A_i is exhausted, then add one element at a time until the complement of A_i is exhausted.) Because these chains are distinct, and there are $t!$ chains of length $t + 1$ altogether,

$$\sum_{i=1}^{m} \alpha_i!(t - \alpha_i)! \leq t!,$$

Dividing this inequality by $t!$ we obtain

$$\sum_{i=1}^{m} \binom{t}{\alpha_i}^{-1} \leq 1.$$

And recognizing that $\binom{t}{k}$ is maximized when $k = \lfloor t/2 \rfloor$, we obtain

$$m\binom{t}{\lfloor t/2 \rfloor}^{-1} \leq \sum_{i=1}^{m} \binom{t}{\alpha_i}^{-1} \tag{3.1}$$

$$\leq 1.$$

Therefore, $m \leq \binom{t}{\lfloor t/2 \rfloor}$. $\square$

What are the possible antichains of $\mathcal{N}_t$ with $\binom{t}{\lfloor t/2 \rfloor}$ elements? Equality in (3.1) can be attained only if $\binom{t}{\lfloor t/2 \rfloor}^{-1} = \binom{t}{\alpha_i}^{-1}$ for each α_i. If t is even, this forces $\alpha_i = t/2$. If t is odd, then α_i can equal $(t-1)/2$ or $(t+1)/2$. We now prove that if t is odd, then all elements of the antichain are size $(t-1)/2$ or all are size $(t+1)/2$. The proof is essentially due to L. Lovász.

Theorem 3.8. *Let A be an antichain of $\mathcal{N}_t$ containing $\binom{t}{\lfloor t/2 \rfloor}$ elements. If t is even, then A is the collection of all subsets of $\mathcal{N}_t$ of size $t/2$. If t is odd, then A is the collection of all subsets of size $(t-1)/2$ or the collection of all subsets of size $(t+1)/2$.*

   Proof.   We have already demonstrated the t even case. Suppose $t = 2u + 1$. As each maximal chain in $\mathcal{N}_t$ contains exactly one element of A, if U is a subset of size u, V is a subset of size $u + 1$, and $U \subseteq V$, then A contains exactly one of U and V. Suppose U is a subset of size u contained in A, and suppose that U' is any other subset of size u. Then there is a sequence of subsets

$$U = U_1, V_1, U_2, V_2, \ldots, V_{n-1}, U_n = U'$$

beginning with U and ending with U', whose sizes alternate between u and $u + 1$, and such that, for each i, V_i contains U_i and U_{i+1}. It follows that U' is an element of A. Because U' was arbitrary, it follows that A contains all subsets of size u. A similar argument shows that if A contains at least one subset of size $u + 1$, then A contains every subset of size $u + 1$. $\square$

We are now ready to look at relations which aren't transitive. In Chapter 4 we begin by discussing graphs, where the relations are merely reflexive and symmetric. The theorems are more difficult to prove in this more general setting — and the analysis of best possible results is *much* more difficult.

Notes

The Erdös–Szekeres theorem was proved in 1935 and may be regarded as a sort of proto-Ramsey theorem (even though Ramsey's theorem was proved in 1930).

We observed that the numbers $1, \ldots, 9$ may be arranged in a sequence with no monotonic subsequence of length 4. How many ways can this be done? Very briefly, each such permutation of $1, \ldots, 9$ corresponds, via the famous *Robinson–Schensted* correspondence, to an ordered pair of standardized Young tableaux of shape $3 + 3 + 3$ (fillings of a 3×3 grid with the numbers $1, \ldots, 9$ so that the numbers increase in each row and column). The so-called *hook-length* formula calculates the number of such tableaux, which in this case turns out to be 42. Therefore the number of desired sequences is $42^2 = 1764$. For a discussion of Young tableaux and the hook-length formula the reader is referred to Chapter 7 and to Van Lint and Wilson (1992).

Exercises

3.1 Find a formula for the number of possible relations on a set of n elements.

3.2 Give an example of a partial order on $|X| = mn$ elements with length m and width n.

3.3 Prove Dilworth's lemma using Theorem 3.4.

3.4 Prove Theorem 3.6.

3.5 Let $n^2 + 1$ distinct points be given in $\mathcal{R}^2$. Prove that there is a sequence of $n + 1$ points $(x_1, y_1), \ldots, (x_{n+1}, y_{n+1})$ for which $x_1 \leq x_2 \leq \cdots \leq x_{n+1}$ and $y_1 \geq y_2 \geq \cdots \geq y_{n+1}$, or a sequence of $n + 1$ points for which $x_1 \leq x_2 \leq \cdots \leq x_{n+1}$ and $y_1 \leq y_2 \leq \cdots \leq y_{n+1}$.

3.6   (Putnam Competition, 1967) Let $0 < a_1 < a_2 < \cdots < a_{mn+1}$ be $mn + 1$ integers. Prove that you can select either $m + 1$ of them no one of which divides any other, or $n + 1$ of them each dividing the following one.

[Hint: Apply Dilworth's lemma.]

3.7   For any $n^2 + 1$ closed interval of $\mathcal{R}$, prove that $n + 1$ of the intervals share a point or $n + 1$ of the intervals are disjoint.

3.8   Prove the infinitary version of the Erdös–Szekeres theorem (Theorem 3.2) as a corollary of the infinitary version of Dilworth's lemma (Theorem 3.6).

3.9   Let $a_1, \ldots, a_n,\ b \in \mathcal{R}$, with all $a_i \geq 1$. Show that the maximum number of sums $\pm a_1 \pm a_2 \pm \cdots \pm a_n$ in the open interval $(b, b + 2)$ is $\binom{n}{\lfloor n/2 \rfloor}$.

3.10 Let $a_1, \ldots, a_n$ be positive real numbers. Show that the maximum number of equal sums $\epsilon_1 a_1 + \cdots + \epsilon_n a_n$ ($\epsilon_i = 0$ or 1) is $\binom{n}{\lfloor n/2 \rfloor}$. See Anderson (1987) and Stanton and White (1986) for a discussion of the connection between the results of Exercises 3.9 and 3.10 and the Littlewood–Offord problem concerning the number of sums $\sum_{i=1}^{n} \epsilon_i z_i$ ($\epsilon_i = \pm 1$ and $|z_i| \geq 1$) lying inside any given circle in the complex plane.

3.11 Let P_1 and P_2 be two total orders on a set of size $n^2 + 1$. Show that there is a subset of size $n + 1$ on which P_1 and P_2 totally agree or totally disagree.

3.12 Show that if $2n - 1$ total orders are given on $m^{2^{2n-1}}$ points, then some $m + 1$ points are totally ordered by n agreeing orders.

4

Ramsey Theory

The Erdös–Szekeres theorem and Dilworth's lemma of Chapter 3 guarantee the existence of particular substructures of certain combinatorial configurations. In other words, they say that large random systems contain nonrandom subsystems. We continue this theme by presenting two cornerstones of Ramsey theory: Ramsey's theorem on graphs and Van der Waerden's theorem on arithmetic progressions. In the process we discuss many related results, including Schur's lemma on equations. We also investigate bounds and asymptotics of Ramsey numbers using techniques from number theory and probability.

4.1. RAMSEY'S THEOREM

The following problem appeared as a question in the 1953 William Lowell Putnam Mathematical Competition:

Problem. *Six points are in general position in space (no three in a line, no four in a plane). The fifteen line segments joining them in pairs are drawn and then painted, some segments red, some blue. Prove that some triangle has all its sides the same color.*

The description of the six points in general position and the segments joining them in pairs is just another way of defining the graph K_6. We introduce a few more graph theory terms. A *coloring* of the set of edges of a graph G (or, loosely, of G itself) is a function $f : E(G) \longrightarrow S$, where S is a set of *colors*. A coloring partitions $E(G)$ into *color classes*. If f is constant, then G is *monochromatic*.

Now we may rephrase the Putnam question as follows: If each edge of K_6 is colored either red or blue, then there is a monochromatic subgraph K_3 (a triangle). We note that the coloring may be done in an arbitrary manner. In fact, because K_6 has $\binom{6}{2} = 15$ edges, there are 2^{15} possible red–blue colorings of the edges of K_6. The claim is that every one of these 32,768 colorings yields a monochromatic K_3. (We are assuming that the vertices of K_6 are labeled, so that we can distinguish between differently labeled isomorphic graphs; and we are also assuming that all 15 edges can be the same color, a possibility that, strictly speaking, is disallowed in the Putnam problem as stated.)

Here is a simple solution to the problem using the pigeonhole principle. Choose any vertex v of K_6. By the pigeonhole principle (Theorem 2.2(3)), some three of the five edges emanating from v are the same color. Without loss of generality, suppose v is connected by red edges to vertices x, y, z. If any of the edges xy, yz, or xz is red, then there is a red triangle (vxy, vyz, or vxz). However, if each of these edges is blue, then xyz is a blue triangle.

A special notation has been introduced to record results of this type. We write

$$K_6 \xrightarrow{2} K_3$$

to indicate that every 2-coloring of the edges of K_6 yields a monochromatic K_3. Similarly, we write

$$K_5 \xnrightarrow{2} K_3$$

to say that there is a 2-coloring of K_5 with no monochromatic K_3. It is equivalent to say that there is a graph G on 5 vertices such that neither G nor G^c contains a K_3. Such a graph is exhibited in Figure 4.1.

In general, we write

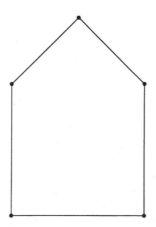

Figure 4.1.   A graph G such that neither G nor G^c contains K_3.

$$K_n \underset{2}{\rightarrow} K_m$$

to indicate that every 2-coloring of the edges of K_n yields a monochromatic K_m. In 1930 F. Ramsey established the existence of such a K_n for each m. Unlike the authors of the Putnam problem, we prefer green–red colorings to red–blue colorings.

Theorem 4.1. *(Ramsey's theorem, 1930). Given $a, b \geq 2$, there exists a least integer $R(a, b)$ with the following property: Every green–red coloring of the edges of the complete graph on $R(a, b)$ vertices yields a green K_a or a red K_b. Furthermore, $R(a, b) \leq R(a - 1, b) + R(a, b - 1)$ for all $a, b \geq 3$.*

Proof. We employ induction on a and b. The basis of the induction consists of the statements $R(a, 2) = a$ and $R(2, b) = b$. These are trivial. In the first assertion, if we two-color K_a and any edge is red, then we obtain a red K_2, while if no edge is red, then we obtain a green K_a. Thus $R(a, 2) \leq a$. Equality follows from the fact that an all green colored K_{a-1} contains neither a green K_a nor a red K_2. The second assertion is proved similarly. Now, assuming the existence of $R(a - 1, b)$ and $R(a, b - 1)$, we will show that $R(a, b)$ exists. Let G be the complete graph on $R(a - 1, b) + R(a, b - 1)$ vertices, and let v be an arbitrary vertex of G. By the pigeonhole principle (Theorem 2.3), at least $R(a - 1, b)$ green edges or at least $R(a, b - 1)$ red edges emanate from v. Without loss of generality, suppose that v is joined by green edges to a complete subgraph on $R(a - 1, b)$ vertices. By definition of $R(a - 1, b)$, this subgraph must contain a red K_b or a green K_{a-1}. In the latter case, the green K_{a-1} and v, and all the edges between the two, constitute a green K_a. We have shown that G contains a green K_a or a red K_b. Therefore, $R(a, b)$ exists and satisfies $R(a, b) \leq R(a - 1, b) + R(a, b - 1)$. $\square$

The integers $R(a, b)$ are called *Ramsey numbers*. Very few of them have been calculated. The fact that we have determined that the Ramsey numbers exist but we don't know their values illustrates one disadvantage of existential proofs.

By definition, $R(m, m)$ is the least positive integer n for which $K_n \underset{2}{\rightarrow} K_m$. The values $R(m, m)$ are called *diagonal Ramsey numbers* because they appear on the main diagonal of a table of Ramsey numbers (Section 4.2). We know one diagonal Ramsey number already: $R(3, 3) = 6$.

We note that $R(a, b) = R(b, a)$ for all $a, b \geq 2$, as the roles of the two variables a and b are symmetric. Furthermore, we have noted that $R(a, 2) = a$ for all a. We have already proved that $R(3, 3) = 6$, but a second proof is furnished by the two observations just made and the inequality $R(a, b) \leq R(a - 1, b) + R(a, b - 1)$ of Theorem 4.1. Thus, $R(3, 3) \leq R(3, 2) + R(2, 3) = 3 + 3 = 6$. The lower bound $R(3, 3) \geq 5$ is verified by construction as before.

Application.   Suppose the vertices of the five-cycle C_5 are a, b, c, d, e (in cyclic order), and that these vertices represent symbols transmitted over a noisy channel. Adjacent symbols are said to be *confusable*; that is, each is easily mistaken for the other. Nonadjacent symbols are not confusable. Thus, c and d are confusable while c and e are not. The independence number $\alpha(G)$ is the maximum number of nonconfusable symbols in $V(G)$. It is easy to see that $\alpha(C_5) = 2$, as illustrated by the pair a, d. For two finite graphs G and H, we define a new graph called the *strong product $G * H$* as the set $V(G) \times V(H)$, with (g, h) adjacent to (g', h') if and only if g is adjacent to or equal to g', and h is adjacent to or equal to h'. We think of $\alpha(G * H)$ as the maximum number of nonconfusable ordered pairs in the set $V(G) \times V(H)$, where nonconfusability means nonconfusability in at least one coordinate. (We are assuming that a symbol is confusable with itself.)

If A is a nonconfusable subset of $V(G)$ and B is a nonconfusable subset of $V(H)$, then $A \times B$ is a nonconfusable subset of $V(G) \times V(H)$. Therefore, $\alpha(G)\alpha(H) \leq \alpha(G * H)$. Ramsey's theorem furnishes a strict upper bound: $\alpha(G * H) < R(\alpha(G) + 1, \alpha(H) + 1)$. For suppose the upper bound is attained by a subset $A \times B$ of $V(G) \times V(H)$. Color an edge green if there is nonconfusability in the first coordinate and red if there is nonconfusability in the second coordinate. (If nonconfusability holds in both coordinates, then we color the edge green.) Ramsey's theorem guarantees that A has at least $\alpha(G) + 1$ points or that B has at least $\alpha(H) + 1$ points – both contradictions. Putting our lower and upper bounds together we obtain $4 = \alpha(C_5)^2 \leq \alpha(C_5 * C_5) < R(\alpha(C_5) + 1, \alpha(C_5) + 1) = R(3, 3) = 6$. Is the value of $\alpha(C_5 * C_5)$ 4 or 5? See Exercise 4.1.

4.2.   GENERALIZATIONS OF RAMSEY'S THEOREM

What we can do with two colors, we can do with an arbitrary number, as the following generalization of Ramsey's theorem shows. All the theorems of this section were proved by F. Ramsey in his 1930 paper. See Notes.

Theorem 4.2. *(Ramsey's theorem for c colors). For any $c \geq 2$ and $a_1, \ldots, a_c \geq 2$, there exists a least integer $R(a_1, \ldots, a_c)$ with the following property: If the edges of the complete graph on $R(a_1, \ldots, a_c)$ vertices are partitioned into color classes $A_1, \ldots, A_c$, then for some i there exists a complete graph on a_i vertices all of whose edges are color A_i.*

Proof. The case $c = 2$ is covered by Theorem 4.1. Suppose $R(a_1, \ldots, a_{c-1})$ exists for all $a_1, \ldots, a_{c-1} \geq 2$. We claim $R(a_1, \ldots, a_c)$ exists and satisfies $R(a_1, \ldots, a_c) \leq R(R(a_1, \ldots, a_{c-1}), a_c)$. A c-coloring of the complete graph on $R(R(a_1, \ldots, a_{c-1}), a_c)$ vertices may be regarded as a 2-coloring with colors $\{A_1, \ldots, A_{c-1}\}$ and A_c. Such a coloring contains a complete graph on a_c vertices colored A_c or a $(c-1)$-colored complete graph on $R(a_1, \ldots, a_{c-1})$ vertices, in which case the induction hypothesis holds. In either case, we obtain a complete subgraph on the appropriate number of vertices. $\square$

Application. We use Theorem 4.2 to prove a weak version of Dilworth's lemma (Theorem 3.3). Recall that Dilworth's lemma states that every partial order on $mn + 1$ elements contains a chain of length $m + 1$ or an antichain of size $n + 1$. For k sufficiently large, we define a coloring of the complete graph on the vertex set $X = \{x_1, \ldots, x_k\}$ as follows: Assuming $i < j$, color edge $x_i x_j$ blue if x_i and x_j are incomparable; green if $x_i \leq x_j$; and red if $x_i \geq x_j$. (Some edges may be colored in two ways, but it won't matter.) Now if $k = R(n + 1, m + 1, m + 1)$, then Theorem 4.2 guarantees a blue K_{n+1} (corresponding to an antichain of size $n + 1$), a green K_{m+1} (corresponding to a chain of x_i with increasing subscripts), or a red K_{m+1} (corresponding to a chain of x_i with decreasing subscripts). Thus we have a weak version of Dilworth's lemma. It is true that the best possible value $mn + 1$ has been replaced by the presumably much larger value $R(n + 1, m + 1, m + 1)$. However, we have gained information about the increasing or decreasing nature of the subscripts of the x_i. It would be unreasonable to expect that the best possible value $mn + 1$ would be obtained by the previous proof, because Dilworth's lemma assumes a transitive relation while Ramsey's theorem does not. Thus Ramsey's theorem is more general than Dilworth's lemma.

We write $K_n \underset{c}{\rightarrow} K_m$ to indicate that every c-coloring of K_n yields a monochromatic K_m. The c-color Ramsey numbers $R(a_1, \ldots, a_c)$ satisfy certain trivial relations, e.g., they are symmetric in the c variables. Furthermore, $R(a_1, \ldots, a_{c-1}, 2) = R(a_1, \ldots, a_{c-1})$ for all a_i, because either there is an edge colored A_c or else all edges are colored from the set $\{A_1, \ldots, A_{c-1}\}$.

A *hypergraph H* of *order n* is a collection of nonempty subsets of an n-set S of *vertices*. The elements of H are called *edges*, corresponding to the graphical case in which edges are two-element subsets. A hypergraph is *t-uniform* if all edges have cardinality t. The *complete t*-uniform hypergraph of order n is the collection $[S]^t$ of all t-subsets of S. One may visualize and draw hypergraphs with edges represented by ovals (cardinality > 2), lines (cardinality $= 2$), and circles (cardinality $= 1$), as in Figure 4.2.

It is now possible to state the most general version of Ramsey's theorem in its natural hypergraph setting.

Theorem 4.3. *(Ramsey's theorem for hypergraphs). Let $c \geq 2$ and $a_1, \ldots, a_c \geq t \geq 2$. There exists a least integer $R(a_1, \ldots, a_c; t)$ with the following property: Every c-coloring of the complete t-uniform hypergraph $[R(a_1, \ldots, a_c; t)]^t$ with colors $A_1, \ldots, A_c$ yields a complete t-uniform hypergraph on a_i vertices in color A_i, for some i.*

Proof. The generalization from 2-colorings to c-colorings works as it did in the generalization from Theorem 4.1 to Theorem 4.2. We leave the argument as an exercise. We know that $R(a_1, a_2; 2)$ exists for all $a_1, a_2 \geq 2$ (Theorem 4.1). Let us assume that $R(a_1, a_2; t-1)$ exists for all $a_1, a_2 \geq 2$, and that $R(a_1 - 1, a_2; t)$ and $R(a_1, a_2 - 1; t)$ exist. We claim that $R(a_1, a_2; t)$ exists and satisfies $R(a_1, a_2; t) \leq n$, where $n = 1 + R(R(a_1 - 1, a_2; t), R(a_1, a_2 - 1; t); t - 1)$. Suppose $[\mathcal{N}_n]^t$ has been

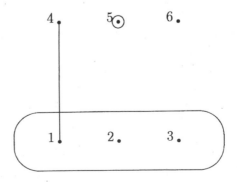

Figure 4.2. A hypergraph with edges $\{1, 2, 3\}$, $\{1, 4\}$, and $\{5\}$.

green–red colored, and let v be one of its vertices. We generate an *induced* 2-coloring of $[\mathcal{N}_n - \{v\}]^{t-1}$ by assigning to each $(t-1)$-set A of $\mathcal{N}_n - \{v\}$ the color which has been assigned to the t-set $A \cup \{v\}$. By definition of n, $[\mathcal{N}_n - \{v\}]^{t-1}$ contains a green $[R(a_1 - 1, a_2; t)]^{t-1}$ or a red $[R(a_1, a_2 - 1; t)]^{t-1}$. Without loss of generality, suppose there is a green $[R(a_1 - 1, a_2; t)]^{t-1}$ on vertex set A. By definition, $[A]^t$ contains a red $[a_2]^t$ or a green $[a_1 - 1]^t$. In the latter case, $[A \cup \{v\}]^t$ contains a green $[a_1]^t$. We have shown that $[\mathcal{N}_n]^t$ contains a green $[a_1]^t$ or a red $[a_2]^t$, as required. $\square$

An application of the theorem to convex sets occurs in the exercises. Now we discuss infinitary versions of Theorems 4.1 and 4.2. We write

$$K_\infty \underset{c}{\rightarrow} K_\infty$$

to indicate that every c-coloring of the complete countably infinite graph yields a monochromatic complete countably infinite subgraph. Similarly, in the hypergraph setting, the *t-uniform complete infinite graph* $K_\infty^{(t)}$ consists of a countably infinite set and all possible t-element subsets. We write

$$K_\infty^{(t)} \underset{c}{\rightarrow} K_\infty^{(t)}$$

to indicate that every c-coloring of the t-uniform complete infinite graph yields a monochromatic t-uniform complete infinite subgraph.

Theorem 4.4. *(Ramsey's theorem for infinite graphs). For every $c \geq 2$,*

$$K_\infty \underset{c}{\rightarrow} K_\infty.$$

Theorem 4.5. *(Ramsey's theorem for infinite hypergraphs). For every $t \geq 2$, $c \geq 2$,*

$$K_\infty^{(t)} \underset{c}{\rightarrow} K_\infty^{(t)}.$$

Theorem 4.5 includes Theorem 4.4 as a special case (when $t = 2$). We now prove Theorem 4.4. The proof of Theorem 4.5 is left as an exercise.

Proof of Theorem 4.4. Define $f : \mathcal{N} \longrightarrow \{1, \ldots, c\}$ as follows. Let $n = 1$ and $X_n = V(K_\infty)$. Choose $x_n \in X_n$ and let $A_i = \{v \in X_n : \text{edge } vx_n$ is color $i\}$. By the infinitary pigeonhole principle (Theorem 2.6), some A_i is infinite. Let $X_{n+1} = A_i$ and define $f(n) = i$ accordingly. Replace n by $n + 1$ and repeat this process.

This recursive procedure defines the function f. Some $f^{-1}(j)$ is infinite (Theorem 2.6), and the complete graph on vertex set $\{x_n : n \in f^{-1}(j)\}$ is monochromatic. $\square$

We close this section by indicating how Ramsey's theorem for infinite graphs implies Ramsey's theorem for finite graphs. The technique for doing this, the so-called "compactness principle," is exploited throughout combinatorics. Assuming the truth of Theorem 4.4, we prove Theorem 4.1 by contradiction. Suppose that there exists a k for which $R(k, k)$ does not exist. For each $i \geq k$, let f_i be a 2-coloring of K_i without a monochromatic K_k. We assume the K_i are nested: $K_k \subseteq K_{k+1} \subseteq K_{k+2} \subseteq \cdots$. By the infinitary pigeonhole principle, there exists an infinite subset of functions $\{f_i^k\} \subseteq \{f_i\}$ which agree on K_k. Similarly, there is an infinite subset of functions $\{f_i^{k+1}\} \subseteq \{f_i^k\}$ agreeing on K_{k+1}, etc. This process yields an infinite 2-coloring of K_∞ without a monochromatic K_k, contradicting Theorem 4.4. Therefore, $R(k, k)$ exists for each k, which implies that $R(a_1, a_2)$ exists, since it surely satisfies the inequality $R(a_1, a_2) \leq R(\max\{a_1, a_2\}, \max\{a_1, a_2\})$. In the same manner, Theorem 4.2 is proven from Theorem 4.4 and Theorem 4.3 is proven from Theorem 4.5.

4.3. RAMSEY NUMBERS, BOUNDS, AND ASYMPTOTICS

Until now our results have been purely existential. We have shown that sufficiently large structures contain desired nonrandom substructures. But how large is sufficiently large? In general, this quantification question is extremely difficult, and unsolved problems abound. We present a few calculations and proofs in this section and summarize the scant amount of information that is known about Ramsey numbers.

We have already shown that $R(3, 3) = 6$. Let us try to evaluate the next more complicated Ramsey number, $R(3, 4)$. To obtain an upper bound, we use the inequality $R(a, b) \leq R(a - 1, b) + R(a, b - 1)$ of Theorem 4.1. Thus $R(3, 4) \leq R(3, 3) + R(2, 4) = 6 + 4 = 10$. However, the true value of $R(3, 4)$ turns out to be 9, not 10. For suppose there is a green–red coloring of K_9 which has no green K_3 and no red K_4. Because $R(2, 4) = 4$ and $R(3, 3) = 6$, it follows that each vertex must be incident with exactly 3 green edges and 5 red edges. But this means that the sum of the degrees of the vertices of the green subgraph is $9 \cdot 3 = 27$, contradicting the fact that the sum of degrees is always even (the handshake theorem). Hence, $R(3, 4) \leq 9$. In the exercises, the reader is asked to furnish a 2-coloring of K_8 containing no green K_3 and no red K_4, thereby proving $R(3, 4) = 9$.

The Ramsey number $R(3, 5)$ is evaluated easily: $R(3, 5) \leq R(3, 4) + R(2, 5) = 9 + 5 = 14$. In the exercises, the reader is asked to find a 2-coloring of K_{13} that shows $R(3, 5) > 13$, thus establishing $R(3, 5) = 14$.

Next we determine the Ramsey number $R(4, 4)$. The inequality $R(a, b) \leq R(a, b - 1) + R(a - 1, b)$ yields the upper bound $R(4, 4) \leq R(4, 3) + R(3, 4) = 9 + 9 = 18$, and 18 turns out to be the value of $R(4, 4)$. To prove this, we need a green–red coloring of K_{17} containing no monochromatic K_4. In general, colorings which establish lower bounds tend to look locally random. However, they must contain quite a bit of structure so that they can be manipulated and analyzed. Such *pseudo-random* constructions are employed throughout combinatorics.

Let us assume that the vertices of K_{17} are labeled with the residue classes modulo 17: 0,1,2,...,16. An edge ij is colored green or red according to the *quadratic character* of $i - j$ modulo 17. The 16 nonzero residues fall into two classes, quadratic residues and quadratic nonresidues. The set of quadratic residues modulo 17 is

$$QR = \{x^2 : x \in \mathcal{Z}_{17}^*\} = \{1, 4, 9, 16, 8, 2, 15, 13\},$$

and the set of quadratic nonresidues is

$$QN = \{3, 5, 6, 7, 10, 11, 12, 14\}.$$

Recall that QR is the range of the homomorphism $f : \mathcal{Z}_{17}^* \longrightarrow \mathcal{Z}_{17}^*$, $f(x) = x^2$. Both QR and QN are closed under multiplication by -1 (because $-1 = 16 \in QR$), so that $i - j$ has the same quadratic character as $j - i$. Edge ij is colored green if $i - j \in QR$ and red if $i - j \in QN$. Suppose there is a monochromatic K_4 on vertices a, b, c, d. First we note that the coloring is translation invariant: $(i + k) - (j + k) = i - j$. Therefore we may assume that $a = 0$. Now multiply each vertex by b^{-1} (the multiplicative inverse of b), and note that either no edge changes color (if $b \in QR$) or else every edge changes color (if $b \in QN$). The reason for this is that $b^{-1}i - b^{-1}j = b^{-1}(i - j)$. In either case, we now have a monochromatic K_4 on vertices 0, 1, cb^{-1}, db^{-1}.

Now, because 1 is a quadratic residue, it follows that the other differences cb^{-1}, db^{-1}, $cb^{-1} - 1$, $db^{-1} - 1$, $db^{-1} - cb^{-1}$ must be quadratic residues. Upon inspection of the elements of QR, we see that this is impossible. Therefore, $R(4, 4) > 17$, and we conclude that $R(4, 4) = 18$.

The other two-color Ramsey numbers $R(a, b)$ are considerably more difficult to evaluate. The above construction involving quadratic residues was discovered in 1955 by R. E. Greenwood and A. M. Gleason. Although it gives the exact Ramsey number in the case of $R(4, 4)$, the method only gives bounds for higher numbers. For example, using this technique one can show that $38 \le R(5, 5)$, but in fact other techniques have shown that $43 \le R(5, 5)$. We present all of the known nontrivial Ramsey numbers in Table 4.1. The notation a/b means that a and b are the best-known lower and upper bounds for that particular Ramsey number. We refer interested readers to Graham, Rothschild, and Spencer (1990).

Open Problem 5. *Determine $R(5, 5)$.*

Table 4.1   Ramsey numbers $R(a, b)$, with $b \ge a$.

a	b							
	3	4	5	6	7	8	9	10
3	6	9	14	18	23	28	36	39/44
4		18	25	43/44				
5			43/49	57/94				
6				102/169				

Open Problem 6. *Determine a formula for $R(n, n)$.*

We know that $R(5, 5) \le R(4, 5) + R(5, 4) = 50$. Unfortunately, this still leaves us with an enormous computation problem in evaluating $R(5, 5)$. The naive approach, examining, say, all $2^{\binom{49}{2}}$ labeled graphs on 49 vertices, is completely intractable.

When we consider more than two colors, the only known nontrivial Ramsey number is $R(3, 3, 3) = 17$, whose proof we leave as an exercise. The only known nontrivial t-uniform hypergraph Ramsey number with $t \ge 3$ is $R(4, 4; 3) = 13$. This state of limited knowledge is exasperating because Ramsey numbers are intimately connected with other numbers and functions, as we shall see later in this chapter. Any new Ramsey number would be very valuable.

Let us now consider lower and upper bounds for the diagonal Ramsey numbers $R(a, a)$. The trivial lower bound $R(a, a) > (a - 1)^2$ is immediate, as one may join with green edges $a - 1$ disjoint copies of a red K_{a-1}, and this coloring has no monochromatic K_a. A more sophisticated lower bound is obtained by the probabilistic method in Section 4.4.

To find an upper bound, we turn to enumerative combinatorics. We recall that $R(a, 2) = a$ for all $a \geq 2$, and $R(a, b) \leq R(a - 1, b) + R(a, b - 1)$ for all $a, b \geq 3$. There is a happy similarity between these laws and the initial condition $\binom{x}{1} = x$ and recurrence relation $\binom{x}{y} = \binom{x-1}{y-1} + \binom{x-1}{y}$ for binomial coefficients.

Theorem 4.6. *For all* $a, b \geq 2$, $R(a, b) \leq \binom{a+b-2}{a-1}$.

Proof. We use induction on a and b, noting that $R(a, 2) = a = \binom{a}{a-1}$ and $R(2, b) = b = \binom{b}{b-1}$, so that the upper bound holds when $b = 2$ or $a = 2$. Suppose the upper bound holds for $R(a-1, b)$ and $R(a, b-1)$, for $a, b \geq 3$. Then

$$R(a, b) \leq R(a - 1, b) + R(a, b - 1)$$
$$\leq \binom{a+b-3}{a-2} + \binom{a+b-3}{a-1}$$
$$= \binom{a+b-2}{a-1},$$

and the upper bound is established. $\square$

For diagonal Ramsey numbers, the inequality of Theorem 4.6 becomes $R(a, a) \leq \binom{2a-2}{a-1}$, and we can determine a nice asymptotic estimate. One of the great open problems of Ramsey theory is to calculate $\lim_{a \to \infty} R(a, a)^{1/a}$ (if it exists).

From Theorem 4.6, it follows that

$$R(a, a) \leq \binom{2a-2}{a-1}$$
$$< 2^{2a-2}$$
$$= 4^{a-1}$$

Thus we obtain an asymptotic upper bound for $R(a, a)^{1/a}$:

$$\limsup R(a,a)^{1/a} \leq \limsup 4^{(a-1)/a}$$
$$= 4.$$

Using Stirling's approximation to $n!$,

$$n! \sim n^n e^{-n}(2\pi n)^{\frac{1}{2}},$$

we can improve the upper bound a little. Since $\binom{2a-2}{a-1} < \binom{2a}{a}$, and

$$\binom{2a}{a} \sim \frac{(2a)^{2a}e^{-2a}(2\pi 2a)^{1/2}}{a^{2a}e^{-2a}2\pi a},$$

it follows that

$$R(a,a) < \frac{2^{2a}}{(\pi a)^{1/2}}(1+o(1)),$$

where $o(1)$ is a function of a which tends to 0 as a tends to ∞.

A lower bound for $\liminf R(a,a)^{1/a}$ is determined in the next section.

4.4. THE PROBABILISTIC METHOD

To obtain a good lower bound for $R(a,a)$, we turn to the probabilistic method, a technique used widely throughout existential combinatorics. See Alon and Spencer (1992). The idea is to turn the objects in question (green–red colorings of a graph) into events in a probability space, and to demonstrate that a desired event (a coloring containing no mono-chromatic subgraph of specified size) occurs with positive probability. If D is a set of desired objects in a sample space S, then the probability that a random object is desired equals $\Pr(D) = |D|/|S|$. If we can show that $\Pr(D)$ is positive, then it follows that D is nonempty; hence there *exists* a desired object. Almost always, probabilistic arguments can be framed directly in terms of the cardinalities $|D|$ and $|S|$. However, the probabilistic language has undisputed bookkeeping and conceptual advantages in proving complex theorems. To illustrate the distinction and parallelism between the two points of view, we present two proofs of the following lower bound for $R(a,a)$, one in terms of cardinalities and the other in terms of probabilities.

Theorem 4.7. *If $\binom{n}{a}2^{1-\binom{a}{2}} < 1$, then $n < R(a,a)$.*

Proof 1 (Cardinalities). Because each of the $\binom{n}{2}$ edges of K_n may be colored independently, the number of green–red colorings is $2^{\binom{n}{2}}$. The number of green–red colorings of K_n with a monochromatic K_a is $|\bigcup A_I|$, where A_I is the collection of green–red colorings in which the subgraph I is monochromatic, and I ranges over all possible subgraphs of K_n isomorphic to K_a. We estimate $|\bigcup A_I|$ as follows:

$$\left|\bigcup A_I\right| \le \sum_I |A_I| \tag{4.1}$$

$$= 2\binom{n}{a}2^{\binom{n}{2}-\binom{a}{2}} \tag{4.2}$$

$$< 2^{\binom{n}{2}}. \tag{4.3}$$

Inequality (4.1) is an enumeration estimate proved by induction on the number of terms in the union; it also follows from the inclusion–exclusion principle of Section 6.1. Statement (4.2) follows from the observation that there are $\binom{n}{a}$ copies of K_a inside K_n. Since each K_a is monochromatic, there are 2 choices for the color of its edges. The remaining $\binom{n}{2} - \binom{a}{2}$ edges of K_n are colored green or red arbitrarily. Finally, (4.3) follows directly from the hypothesis of the theorem.

Now, because $|\bigcup A_I|$ is less than the total number of green–red colorings of K_n, there is a coloring which does not contain a monochromatic K_a. Therefore, $R(a,a) > n$. $\square$

Proof 2 (Probabilities). Suppose the edges of K_n are randomly and independently colored green or red. Think of flipping a coin for each edge. If the coin lands heads, then the edge is colored green; if it lands tails, then red. For each subgraph J of K_n isomorphic to K_a, let A_J be the event that J is monochromatic. We calculate, $\Pr(A_J) = \Pr(J$ is green$) + \Pr(J$ is red$) = 2^{-\binom{a}{2}} + 2^{-\binom{a}{2}} = 2^{1-\binom{a}{2}}$. Therefore,

$$\Pr\left(\bigcup A_J\right) \le \sum_J \Pr(A_J) \quad \text{(subadditivity of probabilities)}$$

$$= \binom{n}{a}2^{1-\binom{a}{2}}$$

$$< 1.$$

The complement of $\bigcup A_J$ occurs with positive probability; therefore, there must be a desired configuration – a 2-coloring of K_n with no monochromatic K_a. Again, $R(a,a) > n$. $\square$

Theorem 4.7 contains an implicit lower bound for $R(a, a)$, if we can untangle it. Fix a and let let N be the minimum value of n satisfying $\binom{n}{a} 2^{1 - \binom{a}{2}} \geq 1$. Then

$$
\begin{aligned}
R(a, a) &\geq N \\
&= (N^a)^{1/a} \\
&> (\binom{N}{a} a!)^{1/a} \\
&\geq (2^{\binom{a}{2} - 1} a!)^{1/a} \\
&= 2^{a/2 - 1/2 - 1/a} a!^{1/a}.
\end{aligned}
$$

From Stirling's asymptotic formula for the factorial function, it follows that

$$
R(a, a) > a 2^{a/2} [\frac{1}{e\sqrt{2}} + o(1)],
$$

Finally, we compute

$$
\begin{aligned}
\liminf R(a, a)^{1/a} &\geq \liminf [a 2^{a/2} [\frac{1}{e\sqrt{2}} + o(1)]]^{1/a} \\
&= \sqrt{2}.
\end{aligned}
$$

Combining this lower bound with the upper bound of Section 4.3, we obtain bounds on $\lim_{a \to \infty} R(a, a)^{1/a}$ (if it exists):

$$
\sqrt{2} \leq \liminf R(a, a)^{1/a} \leq \lim R(a, a)^{1/a} \leq \limsup R(a, a)^{1/a} \leq 4.
$$

These are the best bounds known at present.

Open Problem 7. *Determine whether $\lim_{a \to \infty} R(a, a)^{1/a}$ exists and find its value.*

In 1995 J. H. Kim proved the first conclusive result about the growth of $R(n, k)$ for fixed k. He showed that $R(n, 3) \sim \frac{n^2}{\log n}$.

Open Problem 8. *Determine the asymptotic behavior of $R(n, 4)$.*

In Section 4.5, we leave the quantification question and prove a surprising corollary of Ramsey's theorem.

4.5. SCHUR'S LEMMA

We are now ready to develop the theme of finding order in disorder in contexts other than graph theory. In this section, we prove a proposition about equations as a corollary of Ramsey's theorem, and in Section 4.6 we prove Van der Waerden's theorem, an elegant statement about arithmetic progressions.

For $c, n \geq 1$, we consider a function $f : \mathcal{N}_n \longrightarrow \{A_1, \ldots, A_c\}$. As usual, we think of the A_i as colors, and the function f as assigning a color to each integer, thereby partitioning $\mathcal{N}_n$ into color classes. If f restricted to S is a constant function, then S is *monochromatic*. What kinds of monochromatic structures can we find, given that n is large enough compared to c? One answer to this question was provided by I. Schur in 1916.

Theorem 4.8. *(Schur's lemma). For each $c \geq 1$, there exists a least integer $n = S(c)$ with the following property: for any function $f : \mathcal{N}_n \longrightarrow \{A_1, \ldots, A_c\}$, there exists an A_i containing x, y, z (not necessarily distinct) with $x + y = z$. In other words, there is a monochromatic solution to the equation $x + y = z$.*

Proof. Let $m = R(3, \ldots, 3) - 1$, where $R(3, \ldots, 3)$ is the c-color Ramsey number that guarantees a monochromatic triangle. We claim that m has the desired property, and hence $S(c)$ exists and satisfies $S(c) \leq m$. The function $f : \mathcal{N}_n \longrightarrow \{A_1, \ldots, A_c\}$ generates a c-coloring of the complete graph on vertices $1, 2, \ldots, m + 1$ by assigning to edge ij the color that has been assigned to the integer $|i - j|$. The presence of a monochromatic triangle on, say, vertices a, b, c $(a < b < c)$ implies that the equation $x + y = z$ has the monochromatic solution $(b - a) + (c - b) = (c - a)$. $\square$

Although it is considered an important part of Ramsey theory, Schur's lemma was created by Schur in an attempt to prove Fermat's last theorem. See Notes.

The integers $S(c)$ are called c-color *Schur numbers*. It is trivial to observe that $S(1) = 2$ $(1 + 1 = 2)$. We leave it to the reader to check that $S(2) = 5$ and $S(3) = 14$. The only other known Schur number is $S(4) = 45$. Thus there is a general state of ignorance about Schur numbers, although they are linked to the equally mysterious Ramsey numbers by the inequality $S(c) \leq R(3, \ldots, 3) - 1$.

Open Problem 9. *Find the value of $S(5)$.*

Open Problem 10. *Find a formula for $S(n)$.*

4.6. VAN DER WAERDEN'S THEOREM

Schur's lemma (Theorem 4.8) states that any function $f : \mathcal{N}_n \longrightarrow \{A_1, \ldots, A_c\}$ forces a monochromatic solution to the equation $x + y = z$ (whenever n is sufficiently large compared to c). What other monochromatic structures are forced? One direction for generalization is provided by Rado's theorem, which asserts the existence of a monochromatic solution to the equation

$$\alpha_1 x_1 + \alpha_2 x_2 + \cdots + \alpha_n x_n = 0$$

as long as some nontrival subset of the α_i sums to 0. If this condition is met, we say that the equation is *regular*. For example, the equation $2x_1 - 7x_2 + 3x_3 + 4x_4 - 6x_5 = 0$ is regular $(-7 + 3 + 4 = 0)$. Another direction is provided by B. L. van der Waerden's 1927 theorem concerning arithmetic progressions. An *arithmetic progression* of length l (or $l - AP$) is a sequence

$$a, a + d, a + 2d, \ldots, a + (l - 1)d$$

of l numbers (integers, say), each consecutive pair of which differ by a constant number d. For example, the sequence 20, 30, 40, 50, 60, 70 is a $6 - AP$. Van der Waerden's theorem asserts the existence of a monochromatic $l - AP$ when $\mathcal{N}_n$ is partitioned into c classes (and n is sufficiently large with respect to c and l).

Theorem 4.9. *(Van der Waerden's theorem). Given $c \geq 1$ and $l \geq 1$, there exists a least integer $W = W(c, l)$ with the following property: If $\mathcal{N}_W$ is paritioned into c classes $A_1, \ldots, A_c$, then one of the classes A_i contains a monochromatic $l - AP$.*

Proof. There is an inherent complexity in this theorem, and we only sketch the rather complicated inductive proof on c and l. The theorem is trivially true for some ordered pairs c, l, and in these cases we can actually determine the values of $W(c, l)$: $W(1, l) = l, W(c, 1) = 1, W(c, 2) = c + 1$. The first and third of these statements are the basis of the induction. We shall assume the existence of $W(d, l)$ for every d and prove the existence of $W(c, l + 1)$. The reader is encouraged to rough out a table of c and l, and judge whether this game plan would really cover all ordered pairs c, l. We claim that

$W(c, l+1)$ exists and satisfies $W(c, l+1) \leq f(c)$, where f is defined recursively:

$$f(1) = 2W(c, l)$$
$$f(n) = 2W(c^{f(n-1)}, l)f(n-1), \qquad n \geq 2.$$

As in the proof of Ramsey's theorem, we are establishing an existence result by constructing an upper bound. However, the formulas in the upper bound grow too rapidly to furnish much insight into the exact values of $W(c, l)$.

Let us suppose that $\mathcal{N}_{f(c)}$, which we call an $f(c)$-*block*, is c-colored without a monochromatic $l + 1 - AP$, and $\mathcal{N}_{f(c)}$ is partitioned into $f(c)/f(c-1)$ blocks of $f(c-1)$ consecutive integers, which we call $f(c-1)$- *blocks*. Likewise, each $f(c-1)$-block is partitioned into $f(c-1)/f(c-2)$ blocks of $f(c-2)$ consecutive integers, which we call $f(c-2)$-*blocks*. This partitioning happens at each of the c levels, until, at last, each $f(1)$-block is partitioned into $2W(c, l)$ $f(0)$-*blocks* (which are just integers).

By definition of $W(c, l)$, the first half of each $f(1)$-block contains a monochromatic $l - AP$. Here occurs the first leap of inspiration in the proof. The coloring of the elements of an $f(1)$-block *induces* a coloring of the $f(1)$-block itself. That is, we assign one of $c^{f(1)}$ colors to the $f(1)$-block according to the way its elements are c-colored. Because $f(2) = 2W(c^{f(1)}, l)f(1)$, each $f(2)$-block contains $2W(c^{f(1)}, l)$ $f(1)$-blocks, so that, by definition of $W(c^{f(1)}, l)$, the first half of of each $f(2)$-block contains a monochromatic $l - AP$ of $f(1)$-blocks. Similarly, the first half of each $f(3)$-block contains a monochromatic $l - AP$ of $f(2)$-blocks. This construction happens at each level, so that the first half of $\mathcal{N}_{f(c+1)}$ contains a monochromatic $l - AP$ of $f(c)$-blocks. Let us consider only those integers which lie in $l - AP$s at *all* c levels of blocks. We coordinatize each integer as

$$x = (x_1, \ldots, x_c),$$

with $1 \leq x_i \leq l$, where x_i is the position of x in the monochromatic $l - AP$ of the $f(i)$-block in which it resides. All coordinatized integers have the same color, say A_1. Within each $f(1)$-block, the l integers

$$(1, x_2, \ldots, x_c), (2, x_2, \ldots, x_c), \ldots, (l, x_2, \ldots, x_c)$$

constitute a monochromatic $l - AP$. Therefore, the integer $(l + 1, x_2, \ldots, x_c)$ has a color *other than* A_1, say A_2. Furthermore, the factor 2 in the definition of $f(1)$ implies that $(l+1, x_2, \ldots, x_c)$ occurs within the $f(1)$- block. (The 2 is just a technical constant used to stretch the block enough to accommodate the $(l+1)$-st term of an AP; we could have chosen a smaller constant.) Here occurs the second leap of inspiration: the idea of *focusing*. Within an $f(2)$-block, the l integers

$$(l+1, 1, x_3, \ldots, x_c), (l+1, 2, x_3, \ldots, x_c), \ldots, (l+1, l+1, x_3, \ldots, x_c)$$

are a monochromatic $l - AP$ of color A_2. This forces $(l+1, l+1, x_3, \ldots, x_c)$ to be a color other than A_2. However, we can focus a second $l - AP$ on this integer, namely,

$$(1, 1, x_3, \ldots, x_c), (2, 2, x_3, \ldots, x_c), \ldots, (l, l, x_3, \ldots, x_c).$$

Thus, $(l+1, l+1, x_3, \ldots, x_c)$ cannot be color A_1 or A_2; say it is colored A_3. Figure 4.3 illustrates the two focused progressions, representing colors A_1, A_2, A_3 by dots, circles, and an x, respectively. (The dashes represent numbers with undetermined colors.) Continuing this focusing process at each of the c levels, we conclude that $(l+1, l+1, \ldots, l+1)$ can be none of the colors $A_1, \ldots, A_c$ — a contradiction. Therefore, there must be a monochromatic $l + 1 - AP$. $\square$

The values of $W(c, l)$ are called *Van der Waerden numbers*, and, as we remarked in the proof, the inequality $W(c, l+1) \le f(c)$ does little to establish good estimates for them. In fact, the state of knowledge is even worse for Van der Waerden numbers than for Ramsey numbers. The five known nontrivial Van der Waerden numbers are listed in Table 4.2. The proof of one of these values, $W(2, 3) = 9$, is called for in the exercises.

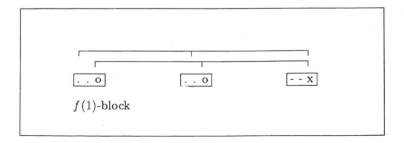

$f(2)$-block

Figure 4.3. Two $3 - AP$s focusing on an integer.

Open Problem 11. *Find* $W(3,4)$.

Although Van der Waerden's theorem asserts the existence of a monochromatic $l - AP$, it does not tell us *which* color it is. The following theorem, whose proof is beyond the scope of this book, guarantees the existence of a monochromatic $l - AP$ in any color that occurs "with positive density." We define the *density function* $d(n, S)$ of a set S of positive integers to be

$$d(n, S) = \frac{|\mathcal{N}_n \cap S|}{n}.$$

The density function measures the fraction of the first n integers which occur in S. Clearly, $0 \le d(n, S) \le 1$ for all n and S.

Theorem 4.10. *(Szemerédi's theorem). For all real numbers $d > 0$ and all $l \ge 1$, there is a positive integer $S(d, l)$ with the following property: If $n \ge S(d, l)$, $S \subseteq \mathcal{N}_n$ and $d(n, S) \ge d$, then S contains an $l - AP$.*

P. Erdós (1913–1996) conjectured the above result in 1935, but it was not proved until 1975 by E. Szemerédi. In 1977 H. Furstenberg gave a proof using ergodic theory.

Conjecture 1.11. *(Erdós). If $\{a_i\} \subseteq \mathcal{N}$ and $\sum 1/a_i$ is a divergent series, then $\{a_i\}$ contains arbitrarily long arithmetic progressions.*

A proof of Erdós' conjecture would settle an outstanding problem about primes. It is well known that $\sum 1/p_i$ diverges if $\{p_i\}$ is the set of primes (see Hardy and Wright (1979)), but it is not known whether there are arbitrarily long arithmetic progressions of primes. A proof of Erdós' conjecture would answer this question affirmatively.

Table 4.2 Van der Waerden number $W(c, l)$, with $c \ge 2, l \ge 3$.

		l		
c		3	4	5
	2	9	35	178
	3	27		
	4	76		

Notes

For original papers of F. Ramsey, P. Erdös and G. Szekeres, and R. P. Dilworth, see Gessel and Rota (1987).

Ramsey numbers have been generalized in many ways. For example, in 1972 V. Chvátal and F. Harary defined the *graph Ramsey number* $r(G, H)$ to be the minimum number of vertices in a complete graph which, when 2-colored, yields a green subgraph G or a red subgraph H. They showed that $r(G, H) > (\chi(G) - 1)(p(H) - 1)$, where $\chi(G)$ is the chromatic number of G and $p(H)$ is the number of vertices of H. They used this inequality to prove $r(T_m, K_n) = (m - 1)(n - 1) + 1$, where T_m is a tree with m vertices. See Graham, Rothschild, and Spencer (1990).

Schur's lemma was proven by I. Schur in an attempt to prove Fermat's last theorem (FLT). Although Schur didn't prove FLT, he did prove that, for all n, if p is prime and sufficiently large, then the congruence $x^n + y^n = z^n$ has a nonzero solution modulo p. Briefly, the argument is to suppose p is prime and greater than $S(n)$. Thus if $\{1, \ldots, p - 1\}$ is n-colored, there exists a monochromatic subset $\{a, b, c\}$ with $a + b = c$. Let $H = \{x^n : x \in \mathcal{Z}_p^*\}$, a subgroup of $\mathcal{Z}_p^*$ of index $\gcd(n, p - 1) \leq n$. The cosets of $\mathcal{Z}_p^*$ define an n-coloring f of $\mathcal{Z}_p^*$ such that $f(a) = f(b) = f(c)$ and $a + b = c$. This implies $1 + a^{-1}b = a^{-1}c$ (in $\mathcal{Z}_p$), and in fact $1, a^{-1}b$, and $a^{-1}c$ are all nth powers in $\mathcal{Z}_p$.

B. L. van der Waerden (1903–1996) proved his 1927 theorem as a generalization of the following conjecture of Schur: If $\mathcal{N}$ is partitioned into two classes, then one of the classes contains arbitrarily long arithmetic progressions.

Ramsey's theorem (in its various formulations) and Van der Waerden's theorem are usually thought of as the two cornerstone theorems of Ramsey theory. See Graham, Rothschild, and Spencer (1990) for a further discussion of these theorems and of other theorems of Ramsey theory, including Gallai's theorem, Rado's theorem, Folkman's theorem, and the Hales–Jewett theorem.

Exercises

4.1 For the confusion graph C_5, show that $\alpha(C_5 * C_5) = 5$.

4.2 A tournament is a complete directed graph. Use Ramsey's theorem (Theorem 4.1) to show that for every n, there exists an $f(n)$

such that every tournament on $f(n)$ vertices contains a transitive subtournament on n vertices.

4.3 (Putnam Competition, 1962) Given five points in a plane, no three of which lie on a straight line, show that some four of these points are the vertices of a convex quadrilateral.

4.4 Use Theorem 4.3 and the result of the previous exercise to prove the following result (1935) of P. Erdös and G. Szekeres: For every m, there exists a least integer $n(m)$ such that any set of $n(m)$ points in the plane contains m points which determine a convex m-gon. Hint: $n(m)$ satisfies $n(m) \leq r(5, m; 4)$. Actually, Erdös and Szekeres proved that $n(m) \geq 2^{m-2} + 1$ and conjectured that $n(m) = 2^{m-2} + 1$. The determination of $n(m)$ remains an open problem.

4.5 Use Theorem 4.4 to show that among infinitely many points in the plane there are infinitely many collinear points or infinitely many points no three of which are collinear. Prove also the three-dimensional analogue of this problem: Among infinitely many points in R^3 there is an infinite planar subset or an infinite subset containing no four coplanar points.

4.6 Prove Theorem 4.5.

4.7 A c-coloring of a graph G which uses all c colors is said to be an *exact c-coloring* of G. For integers $a \geq b \geq 1$, let $P(a, b)$ stand for the following proposition: Every exact a-coloring of the infinite complete graph G yields an exactly b-colored infinite subgraph H. Show that $P(a, b)$ is true if $b = 1$, $b = 2$, or $a = b$. Show that $P(10, 8)$ is false. It is not known what pairs a, b make $P(a, b)$ a true proposition.

4.8 Find a 2-coloring of K_8 which proves $R(3, 4) > 8$.

4.9 Find a 2-coloring of K_{13} which proves $R(3, 5) > 13$.

4.10 Prove $R(3, 3, 3) = 17$.

4.11 Show that if K_{327} is 5-colored, there exists a monochromatic K_3.

4.12 Prove that if $K_{\infty,\infty}$ is 2-colored, there exists a monochromatic $K_{\infty,\infty}$. Interpret this result as a proposition about 2-colorings of the infinite square lattice.

4.13 Use Theorem 4.7 to obtain a lower bound for $R(100, 100)$. (Hint: $3.0038 \cdot 10^{16}$ is possible.)

4.14 Use the probabilistic method to prove that almost all labeled graphs have diameter two; hence almost all labeled graphs are connected.

4.15 Use the probabilistic method to prove Schütte's theorem: For every m there exists a tournament T such that for each $S \subseteq T$, $|S| = m$, there exists a vertex $p \in T - S$ which is directed to each vertex of S. Find such tournaments for $m = 1$ and $m = 2$. Hint: the tournament for $n = 2$ can be constructed from the set of quadratic residues modulo 7 as follows. Let QR and QN be the set of quadratic residues and nonresidues modulo 7, respectively. Put a directed arrow from vertex i to vertex j if $i - j \in QR$ and an arrow from j to i if $i - j \in QN$. Now check that this tournament has the desired property.

[Note. Schütte's theorem was proved by P. Erdös in 1963.]

4.16 Prove $S(2) = 5$ and $S(3) = 14$.

4.17 (Putnam Competition, 1988) (a) If every point of the plane is painted one of three colors, do there necessarily exist two points of the same color exactly one inch apart?

(b) What if "three" is replaced by "nine"?

Justify your answers.

[Note: The answer to (a) is yes and the answer to (b) is no. The minimum number of colors necessary to force the conclusion in part (a) is not known. See Klee and Wagon (1991).]

4.18 (Putnam Competition, 1994) Show that if the points of an isosceles right triangle of side length 1 are each colored with one of four colors, then there must be two points of the same color which are at distance at least $2 - \sqrt{2}$ apart.

4.19 Prove or disprove: If $\mathcal{N}$ is 2-colored, there exists a monochromatic infinite arithmetic progression.

4.20 Prove $W(2, 3) = 9$.

4.21 Find upper bounds for $W(3, 4)$ and $W(4, 4)$.

4.22 Prove or disprove: If $\mathcal{R}$ is 2-colored, then there exist $a, b, c \in \mathcal{R}$ with a, b, c all the same color and $(c - b)/(b - a) = \sqrt{2}$.

4.23 (Putnam Competition, 1960) Consider the arithmetic progression

$$a, a + d, a + 2d, \ldots,$$

where a and d are positive integers. For any positive integer k, prove that the progression has either no exact kth powers or infinitely many.

4.24 Find a positive integer n with the following property: No matter how $\mathcal{N}_n \times \mathcal{N}_n$ is 2-colored, there exist $i, j, k \in \mathcal{N}_n$ such that the set

$$\{(i,j), (i + k, j), (i, j + k), (i + k, j + k)\}$$

is monochromatic. Hint: $n = 9(2^{81} + 1)(2^{(2^{81}+1)^2} + 1)$ suffices.

4.25 Find an infinite graph G with the following three properties:

1. G contains no K_4.
2. The addition of any edge to G completes a K_4.
3. There is a two-coloring of the edges of G with no monochromatic K_3.

In 1988 J. Spencer used the probabilistic method to prove the existence of a finite graph G with properties (1) and (2), *without* property (3), and with fewer than $3 \cdot 10^9$ vertices. His proof answers a question of P. Erdös, who asked whether there exists such a graph with at most 10^{10} vertices. See J. Spencer, Three hundred million points suffice, *Journal of Combinatorial Theory (A)* **49** (1988), 210–217 and Erratum, *Journal of Combinatorial Theory (A)* **50** (1989), 323.

For $m, n \geq 3$ and $p > \max\{m, n\}$, the *Folkman number* $F(m, n; p)$ is defined as the minimum number of vertices in a graph G with the properties (1) G contains no complete subgraph K_p, and (2) any green–red coloring of the edges of G yields a green K_m or a red K_n. In 1967 J. Folkman proved the existence of these numbers. Spencer's construction proves $F(3, 3; 4) < 3 \cdot 10^9$. Other than the Ramsey numbers (i.e., when $p > R(m, n)$), the only known Folkman number is $F(3, 3; 6) = 8$. The only Folkman number which has been reasonably bounded is $F(3, 3; 5)$: $10 \leq F(3, 3; 5) \leq 17$. See M. Erickson, An upper bound for the Folkman number F(3,3;5), *Journal of Graph Theory* **17** (1993), 679–681.

4.26 Find a $5 - AP$ of prime numbers.

4.27 Find a $6 - AP$ of prime numbers.

4.28 Prove that for any positive integers c and l, there exists a number W with the property that, whenever the set $\mathcal{N}_W$ is c-colored, there exists an $l - AP$ with each of its terms and the common difference the same color.

4.29 Using the compactness principle, prove that the following proposition is equivalent to Van der Waerden's theorem: For all $c, l \geq 1$, no matter how $\mathcal{N}$ is c-colored, there exists a monochromatic $l - AP$.

PART

Enumeration

II

The expected number of fixed points in a permutation of n objects (for example, a deck of n cards) is exactly one, independent of n.

The probabilistic formulation of this proposition is a thin veil over a counting problem, and counting problems are notoriously difficult. (Here the problem is to count the total number of fixed points in the group of permutations of n objects.) We give a difficult brute force solution in Chapter 6 before giving two elegant solutions in Chapter 8. One of the elegant solutions is an easy application of Burnside's lemma, which is proven using the important technique of enumerating a set two different ways and equating the results. In Part II we study many kinds of counting problems. As the above example suggests, we seek not only formulas for intricate combinatorial situations, but also connections and unifying principles. We begin in Chapter 5 by overviewing the *fundamental problem of enumeration theory*, which is the problem of counting functions. This problem devolves into sixteen basic cases which we examine in Chapter 6. In the more advanced cases, the function counting problem requires group theory, cycle indexes, and Pólya's theory of counting. These topics are treated in Chapter 8.

5

The Fundamental Counting Problem

Enumeration is probably the trickiest branch of combinatorics. Sadly, it is common to hear students say, "I just can't count," or "I don't count," or "I count, but I always get two different answers." There is confusion about what is being counted and there are many formulas to remember. Although the situation is fraught with difficulty – even desperation – there is a solution. Suprisingly, one general principle and a few variations suffice for nearly all enumeration problems one is likely to encounter. This point of view is not original to this book. The same basic approach (in more depth) is undertaken by Stanley (1986). The idea is to represent the objects to be counted as functions $f : X \longrightarrow Y$, where X and Y are chosen appropriately. The conditions imposed in the enumeration problem usually amount to putting restrictions on the functions (e.g., requiring them to be one-to-one), and/or making rules as to when two functions are considered equivalent (e.g., when they are equivalent up to a permutation of the elements of X). In Chapter 5 we consider this object–function approach to enumeration, and we itemize the sixteen basic cases.

5.1. LABELED AND UNLABELED SETS

Suppose X and Y are two finite nonempty sets and Y^X is the collection of functions $f : X \longrightarrow Y$. We wish to define some equivalence relations on Y^X and count the number of equivalence classes. When we say that two functions f and g ($f, g \in Y^X$) are *equivalent*, we will mean one of four things:

    1. $f = g$.
    2. $f = gh$ for some bijection $h : X \longrightarrow X$.
    3. $f = ig$ for some bijection $i : Y \longrightarrow Y$.

4. $f = igh$ for bijections $h : X \longrightarrow X$ and $i : Y \longrightarrow Y$.

(Note that the functions here are applied from right to left.)

In definitions (2) and (4) we say that h delabels X, and we speak of X as an *unlabeled* (or *delabeled*) set. Likewise, in definitions (3) and (4) we say that i delabels Y, and we speak of Y as an unlabeled or delabeled set.

Example. Consider the functions f and g of Figure 5.1. The functions fail to be equivalent under definition (1) because, after all, they are different functions. However, if $h : X \longrightarrow X$ is the bijection given by $h(a) = a$, $h(b) = c$, and $h(c) = b$, then $f = gh$; therefore f and g satisfy definition (2). The bijection h rearranges the set $\{a, b, c\}$, eliminating the discrepancy between the two functions due to the labeling of the elements of the domain. If i is the bijection given by $i(x) = x$, $i(y) = z$, and $i(z) = y$, then $f = ig$; therefore f and g satisfy definition (3). Now it is clear that the functions f and g of Figure 5.1 are equivalent according to definitions (2), (3), and (4).

Example. The functions f and g of Figure 5.2 are equivalent only when both X and Y are unlabeled, i.e., according to definition (4). Define $h : X \longrightarrow X$ by $h(a) = c$, $h(b) = b$, and $h(c) = a$, and define $i : Y \longrightarrow Y$ by $i(x) = z$, $i(y) = x$, and $i(z) = y$. Then $f = igh$.

There are presumably $2^4 = 16$ possibilities for the equivalence or inequivalence of f and g according to the four definitions. How many of these possibilities can actually occur? Answer: 6. The reader is

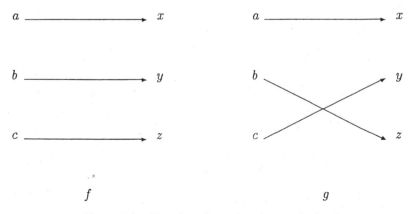

Figure 5.1. Two functions. Are they equivalent?

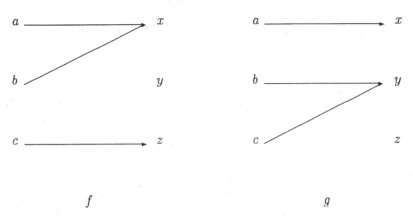

Figure 5.2. Equivalent functions when domain and codomain are unlabeled.

encouraged to construct six illustrative examples. (We have already given two.)

In the next section we impose restrictions on the functions to be counted.

5.2. THE SIXTEEN CASES

We saw in the previous section that the domain and codomain of a function may be labeled or unlabeled sets, leading to four types of functions to be counted. Furthermore, functions may be classified according to whether they are one-to-one, onto, both, or not necessarily either. Altogether there are sixteen cases, and these are organized into Table 5.1.

We assume that $f : X \longrightarrow Y$ is a function from a set X with x elements to a set Y with y elements. Let us consider the entries in the table, beginning with the X labeled, Y labeled box. We repeat some of the analysis done in Section 1.2. For example, we have already noted that the total number of such functions is y^x.

As for the second entry, every one-to-one (1 - 1) function corresponds to an ordered selection of x objects from the set Y. There are y choices for the first object selected, $y - 1$ choices for the second, and so on, leading to the formula

$$P(y, x) = y(y - 1)(y - 2) \ldots (y - x + 1) = \frac{y!}{(y - x)!}$$

Table. 1 The number of functions from X to Y.

X		labeled		unlabeled	
		Y			
	total	y^x	total	$B(x)$	$x \leq y$
	1–1	$p(y, x)$	1–1	1	$x \leq y$
labeled				0	$x < y$
	onto	$T(x, y)$	onto	$S(x, y)$	
	bijections	$x! \delta(x, y)$	bijections	$\delta(x, y)$	
	total	$\binom{x+y-1}{y-1}$	total	$p(x)$	$x \leq y$
	1–1	$\binom{y}{x}$	1–1	1	$x \leq y$
unlabeled				0	$x \leq y$
	onto	$\binom{x-1}{y-1}$	onto	$p(x, y)$	
	bijections	$\delta(x, y)$	bijections	$\delta(x, y)$	

if $x \leq y$. Because there are not enough objects to permute if $x > y$, we decree that $P(y, x) = 0$ for $x > y$.

A more interesting formula arises in the case of onto functions. For now, we merely call the number of such functions $T(x, y)$ and promise to deal with it in Chapter 6.

In the case of bijections, we either have two sets of the same cardinality or we don't. If $x \neq y$, then no bijection is possible. However, if $x = y$, then there are $x!$ ways to match up the two sets. We define $\delta(x, y)$ to be 0 if $x \neq y$ and 1 if $x = y$. Thus the number of bijections is $x! \delta(x, y)$. If either X or Y (or both) is unlabeled, then all bijections look the same, so the total number is $\delta(x, y)$.

The entries in the X unlabeled, Y labeled box are standard arguments of discrete mathematics. A one-to-one function $f : X \longrightarrow Y$ is equivalent to a selection of x elements of Y without regard to order. The number of such selections is the value of the binomial coefficient

$$\binom{y}{x} = \frac{y!}{x!(y - x)!}.$$

A function $f : X \longrightarrow Y$ is a distribution of x identical elements (the elements of X) into y different boxes (the elements of Y). A box may receive any number of objects, including zero. Suppose we represent the x identical objects with x copies of the symbol O. The y boxes are represented by a set of $y - 1$ vertical lines |. The placement of the objects in the boxes is indicated by a linear ordering of the O's and |'s. For example, OOO|O| |O means that the first box (to the left of the first |) contains three objects, the second box contains one object, the third contains no objects, and the fourth contains one object. The total number of items in the linear ordering is $x + y - 1$, and $y - 1$ of these are |'s. Therefore, the total number of functions is $\binom{x+y-1}{y-1}$. If f is onto, then each box must get at least one object, and there are $x - y$ objects to distribute freely. Thus the number of onto functions is $\binom{x-y+y-1}{y-1} = \binom{x-1}{y-1}$. Can the reader furnish a more direct combinatorial proof of this formula?

If X is labeled and Y unlabeled, then the total number of functions from X to Y is the number of ways X may be partitioned into unlabeled parts (corresponding to the images under f). We prefer to define a formula only when $x \leq y$, in which case the magnitude of y becomes unimportant (X can't be divided into more than x parts). The *Bell number* $B(x)$ is the number of such functions. The formula for $B(x)$ is derived in Chapter 6.

If $x > y$, then one-to-one functions do not exist. However, if $x \leq y$, then all one-to-one functions look alike when X is unlabeled. This observation accounts for the 1–1 entries of the X unlabeled, Y labeled box and the X unlabeled, Y unlabeled box.

We define $S(x, y)$ to be the number of onto functions when X is labeled and Y is unlabeled. The values of $S(x, y)$ are called *Stirling numbers* of the *second kind*. (We will define Stirling numbers of the first kind in Chapter 6.) Clearly, $S(x, y) = T(x, y)/y!$, as dividing by $y!$ permutes the labels of the y sets into which X has been divided. Stirling numbers of the second kind are treated in Chapter 6.

The total number of functions when X and Y are unlabeled and $x \leq y$ is denoted $p(x)$, and the values of $p(x)$ are called *partition numbers*. The number of onto functions is $p(x, y)$, and these values are also called partition numbers. Partition numbers are discussed in Chapter 6.

We have calculated, or at least named, all sixteen entries of Table 5.1. We have yet to determine formulas for $T(x, y)$ and $S(x, y)$, as well as for the Bell numbers $B(x)$ and the partition numbers $p(x)$ and $p(x, y)$. We have already mentioned that $S(x, y)$ and $T(x, y)$ are related by the equation $S(x, y) = T(x, y)/y!$. Four more equations are obtained by comparing the total number of functions in each box to the number of onto functions. Thus,

$$y^x = \sum_{j=1}^{x} T(x,j) \binom{y}{j} \tag{5.1}$$

$$\binom{x+y-1}{y-1} = \sum_{j=1}^{x} \binom{x-1}{j-1} \binom{y}{j} \tag{5.2}$$

$$B(x) = \sum_{j=1}^{x} S(x,j) \tag{5.3}$$

$$p(x) = \sum_{j=1}^{x} p(x,j). \tag{5.4}$$

Equation (5.1), for example, follows from the fact that every function $f : X \longrightarrow Y$ is onto *some* subset $Y' \subseteq Y$ of cardinality j, $1 \leq j \leq x$. The binomial coefficient $\binom{y}{j}$ "chooses" Y' and $T(x,j)$ counts the number of functions from X onto Y'. Equations (5.2) through (5.4) are proved similarly.

From these identities and the equations mentioned earlier, it follows that we need only find formulas for $T(x,y)$ and $p(x,y)$. The other formulas can be written in terms of these. The determination of $T(x,y)$, a pretty application of the inclusion–exclusion principle, is the subject of Section 6.1. The calculation of $p(x,y)$ is a much more difficult matter, and we present no explicit formula. We merely content ourselves with producing a generating function for $p(x,y)$ in Section 6.4.

Notes

The organization into sixteen cases is called the *twelvefold way* (bijections are not treated separately) by Stanley (1986).

Exercises

5.1 Verify the formulas for the X unlabeled, Y labeled case in Table 5.1.

5.2 Verify formulas (5.2), (5.3), and (5.4).

5.3 Show that formula (5.2) follows directly from formula (1.5) upon a change of variables.

5.4 How many functions $f : X \longrightarrow X$, $|X| = n$, have the property that $f(f(x)) = x$ for all $x \in X$? See Exercise 8.6.

5.5 How many functions $f : X \longrightarrow Y$ (X unlabeled, Y labeled) have the property that $|f^{-1}(y)|$ is an odd number of all $y \in Y$?

5.6 In how many ways may x indistinguishable balls be placed into y urns so that each urn contains an odd number of balls?

5.7 Give a bijective proof that $\binom{x-1}{y-1}$ is the number of ways of placing x indistinguishable balls into y urns so that each urn is nonempty.

5.8 (a) Find a formula for $a(m,n)$, the number of functions $f : \mathcal{N}_m \longrightarrow \mathcal{N}_n$ with the property that $f(x) < f(y)$ whenever $1 \le x < y \le m$.
(b) Find a formula for $b(m,n)$, the number of functions $f : \mathcal{N}_m \longrightarrow \mathcal{N}_n$ with the property that $f(x) \le f(y)$ whenever $1 \le x < y \le m$.

5.9 (a) Find a formula for $c(m,n)$, the number of solutions (ordered n-tuples of nonnegative integers) to $m = x_1 + \cdots + x_n$.
(b) Find a formula for $d(m,n)$, the number of solutions (ordered n-tuples of positive integers) to $m = x_1 + \cdots + x_n$.
(c) Find a formula for $e(m,n)$, the number of solutions to $m = x_1 + \cdots + x_n$ with $x_1 \le \cdots \le x_n$.
(d) Find a formula for $f(m,n)$, the number of solutions to $m = x_1 + \cdots + x_n$ with $x_1 < \cdots < x_n$.

5.10 Let $S(n) = |\{(n_1,\ldots,n_k) : k, n_i \in \mathcal{N}, \sum_{i=1}^{k} n_i = n\}|$. Determine with proof a formula for $S(n)$. Note that $S(n)$ counts the number of ways n may be written as $n = n_1 + \cdots + n_k$ for any k (order important). Such summations are called *compositions* of n.

6

Recurrence Relations and Explicit Formulas

In this chapter we find formulas for the Stirling numbers of the first and second kind and the Bell numbers, as well as relations satisfied by the partition numbers. Moreover, along the way we develop some general machinery, particularly the inclusion–exclusion principle, recurrence relations, and generating functions.

6.1. THE INCLUSION–EXCLUSION PRINCIPLE

As in Chapter 5, we let $T(x, y)$ be the number of onto functions $f : X \longrightarrow Y$ ($|X| = x$, $|Y| = y$). The machinery necessary to calculate $T(x, y)$ is the inclusion–exclusion principle, a generalization of the familiar Venn diagram rule: $|A \cup B| = |A| + |B| - |A \cap B|$.

Theorem 6.1. *(Inclusion–exclusion principle). For $A_1, \ldots, A_n \subseteq S$, S finite,*

$$|A_1 \cup \cdots \cup A_n| = \sum_{j=1}^{n} (-1)^{j+1} \sum |A_{k_1} \cap \cdots \cap A_{k_j}|, \qquad (6.1)$$

with the second sum over all j-tuples $(k_1, \ldots, k_j)$ with $1 \leq k_1 < \cdots < k_j \leq n$.

Proof. Let $s \in S$ and assume that s is contained in exactly m of the A_i. The contribution of s to the right side of (6.1) is 0 if $m = 0$. If $m \geq 1$, then the contribution is

$$\sum_{j=1}^{n}(-1)^{j+1}\binom{m}{j} = \sum_{j=1}^{m}(-1)^{j+1}\binom{m}{j} \quad \text{(because } m \leq n)$$

$$= (-1)[\sum_{j=0}^{m}\binom{m}{j}(-1)^{j} - 1]$$

$$= (-1)[(-1+1)^{m} - 1]$$

$$= 1.$$

Therefore, each $s \in S$ not in the union of the A_i contributes zero to both sides of (6.1), while each $s \in S$ in the union contributes 1. This means that each element of S contributes an equal amount to both sides of (6.1); hence, (6.1) is a valid equation. $\square$

Application (The "problem of derangements"). A bijection $f : \mathcal{N}_n \longrightarrow \mathcal{N}_n$ with no fixed points is called a *derangement*. What is the probability P_n that a random bijection $f : \mathcal{N}_n \longrightarrow \mathcal{N}_n$ is not a derangement? Simple calculations show that $P_1 = 1$, $P_2 = 1/2$, $P_3 = 2/3$, and $P_4 = 15/24$. To obtain a general formula, let A_i be the collection of functions which have i as a fixed point. The inclusion–exclusion principle says that

$$|A_1 \cup \cdots \cup A_n| = \sum_{j=1}^{n}(-1)^{j+1}\binom{n}{j}(n-j)!.$$

Therefore, $P_n = \sum_{j=1}^{n}\frac{(-1)^{j+1}}{j!}$. At first, it may seem strange that a fixed point is less likely to occur when n is 52 than when n is 51 or 53. It is interesting to note that

$$\lim_{n\to\infty} P_n = 1 - e^{-1} \doteq 0.63.$$

At this point we can give a brute force proof of the gem mentioned at the beginning of Part II: The expected number of fixed points of an element of S_n is 1. We will give two much more elegant proofs in Section 8.1.

Let D_n be the number of derangements in S_n. By the above analysis,

$$D_n = \sum_{j=0}^{n}(-1)^j\frac{n!}{j!}.$$

Thus the expected number of fixed points in an element of S_n is

$$E = \frac{1}{n!} \sum_{k=1}^{n} k \binom{n}{k} D_{n-k}$$

$$= \sum_{k=1}^{n} \frac{k}{k!(n-k)!} \sum_{j=0}^{n-k} (-1)^j \frac{(n-k)!}{j!}$$

$$= \sum_{k=0}^{n-1} \sum_{j=0}^{n-k-1} \frac{(-1)^j}{k!j!}$$

$$= \sum_{t=0}^{n-1} \sum_{k+j=t} \frac{(-1)^j}{k!j!}$$

$$= 1 + \sum_{t=1}^{n-1} \frac{1}{t!} \sum_{j=0}^{t} (-1)^j \binom{t}{j}$$

$$= 1 + \sum_{t=1}^{n-1} \frac{1}{t!} \cdot 0$$

$$= 1.$$

The above computation illustrates the method of evaluating a sum by changing the order of summation. The simplest reordering of a double summation is an interchange of the two sigmas. But an interchange doesn't work here, so we tried something else: summing along diagonals.

Application. Euler's ϕ function is defined for all natural numbers by

$$\phi(n) = |\{1 \le x \le n : \gcd(x, n) = 1\}|.$$

Suppose the canonical factorization of n into prime powers is $n = \prod_{i=1}^{k} p_i^{\alpha_i}$. Let $X_i = \{y : 1 \le y \le n \text{ and } p_i \mid y\}$. Then

$$\phi(n) = n - |X_1 \cup \cdots \cup X_k|$$

$$= n - \left(\frac{n}{p_1} + \frac{n}{p_2} + \cdots\right) + \left(\frac{n}{p_1 p_2} + \frac{n}{p_1 p_3} + \cdots\right) - \cdots$$

$$= n\left(1 - \frac{1}{p_1}\right) \cdots \left(1 - \frac{1}{p_k}\right).$$

Application. To calculate $T(x, y)$, let S be the collection of functions $f : X \longrightarrow Y$, $|X| = x$, $Y = \{1, \ldots, y\}$. We have already observed in Section 5.2 that S has cardinality y^x. We now wish to find the cardinality of the subset of S consisting of all onto functions. Let A_k be the collection of functions whose ranges do not contain k. Then the intersection of j of the A_k has cardinality $(y-j)^x$, the number of unrestricted functions from X to the nonexcluded $y-j$ elements of Y. Applying the inclusion–exclusion principle, we obtain

$$|A_1 \cup \cdots \cup A_y| = \sum_{j=1}^{y} (-1)^{j+1} \binom{y}{j} (y-j)^x.$$

Because $T(x,y)$ is the complement of this union, it follows that

$$T(x,y) = y^x - |A_1 \cup \cdots \cup A_y|$$

$$= y^x - \sum_{j=1}^{y} (-1)^{j+1} \binom{y}{j} (y-j)^x$$

$$= \sum_{j=0}^{y} (-1)^{j} \binom{x}{j} (y-j)^x.$$

Although it contains a summation sign as well as the binomial coefficient $\binom{y}{j}$, the above formula is the most compact expression possible for $T(x,y)$. As what we desire to count becomes more complicated, we must accept more involved formulas. The curious reader may wish to look ahead to Section 8.4 for a very intricate formula for counting graphs. In Section 6.2 we use our formula for onto functions to derive expressions for the Stirling numbers and the Bell numbers.

The inclusion–exclusion principle of Theorem 6.1 can be generalized to the well-known Bonferonni inequalities of probability theory. For each j, $1 \le j \le n$, let

$$N_j = \sum |A_{k_1} \cap \cdots \cap A_{k_j}|,$$

with the sum over all j-tuples vectors $(k_1,\ldots,k_j)$ with $1 \le k_1 < \cdots < k_j \le n$.

Theorem 6.2. *(Bonferonni inequalities). For $A_1,\ldots,A_n \subseteq S$, S finite,*

$$|A_1 \cup \cdots \cup A_n| \le \sum_{j=1}^{t} (-1)^{j+1} N_j,$$

if t is odd. If t is even, the inequality is reversed.

Proof. As in the proof of Theorem 6.1, let $s \in S$ and assume that s is contained in exactly m of the A_i. If $m = 0$, then the contribution to both sides of the inequality is 0. For $m > 0$, the result follows easily from the identity $\sum_{j=0}^{t} \binom{m}{j}(-1)^j = \binom{m-1}{t}(-1)^t$, which was proved in Section 1.3. $\square$

6.2. STIRLING NUMBERS

In Section 5.2 we defined the Stirling number $S(x, y)$ to be the number of onto functions $f : X \longrightarrow Y$, where X is labeled and Y unlabeled ($|X| = x$ and $|Y| = y$). Any such function may be viewed as a partition of X into y unlabeled subsets. As $T(x, y)$ counts the same collection when Y is labeled, it follows that $S(x, y) = T(x, y)/y!$. Thus the formula for $T(x, y)$ from Section 6.1 furnishes an immediate expression for $S(x, y)$:

$$S(x, y) = \frac{1}{y!} \sum_{j=0}^{y} (-1)^j \binom{y}{j} (y - j)^x. \tag{6.2}$$

Equation (6.2) can be simplified somewhat by replacing j by $y - j$:

$$S(x, y) = \frac{1}{y!} \sum_{j=0}^{y} (-1)^{y-j} \binom{y}{j} j^x. \tag{6.3}$$

Equations (6.2) and (6.3) are explicit formulas for the Stirling numbers of the second kind. However, they are difficult to use for calculations. It is much more convenient to use a recurrence procedure, i.e., a set of initial values and a rule for computing higher values from lower ones. Such a procedure is indeed possible for $S(x, y)$. We note that $S(x, 1) = 1$ for all x, as there is only one way to partition X into one subset. Also, $S(x, x) = 1$ for all x, as X may be partitioned into x subsets in only one way. Now let us find a way to compute $S(x, y)$ from previous values. If $X = \{1, \ldots, x\}$, then the element 1 can be alone in a class of the partition, or it can be in a class with other elements. If it is alone, then there are $S(x - 1, y - 1)$ ways to partition $X - \{1\}$ into the other $y - 1$ subsets. However, if 1 is in a class with other elements, then there are y choices for which class contains 1, and $S(x - 1, y)$ ways to partition $X - \{1\}$ into y classes. Thus,

$$S(x,1) = 1, \quad S(x,x) = 1 \quad \text{for all } x. \tag{6.4}$$
$$S(x,y) = S(x-1,y-1) + yS(x-1,y) \quad \text{for all } x \geq 2. \tag{6.5}$$

From (6.4) and (6.5) we immediately obtain a table (Table 6.1) of values of $S(x,y)$ for small x and y. Recalling formula (4.3), we see that the marginal sums of this table are the Bell numbers.

Note that $S(x,y)$ is unimodal for each x (the values increase to a point and then decrease). The unsigned Stirling numbers of the first kind to follow are also unimodal. Let us verify an entry of the table, say $S(4,3) = 6$. We need to show that there are 6 ways to partition the set $\{1,2,3,4\}$ into three subsets. By inspection, we list the different partitions (suppressing commas and one level of set notation): $\{12,3,4\}$, $\{1,3,24\}$, $\{1,2,34\}$, $\{13,2,4\}$, $\{1,4,23\}$, $\{14,2,3\}$.

The existence of Stirling numbers of the second kind may lead one to wonder whether there are Stirling numbers of the first kind. There are, and the two types of Stirling numbers are intimately linked by generating functions and by an inversion formula.

Unlike the Stirling numbers of the second kind, which are all positive, the Stirling numbers of the first kind are signed. The significance of the sign is discussed in the next section. The *Stirling number of the first kind* $s(n,k)$ is defined to be $(-1)^{n+k}|s(n,k)|$, where $|s(n,k)|$ is the number of permutations in the symmetric group S_n which have k cycles. For example, $|s(3,2)| = 3$, as there are 3 permutations in S_3 with two cycles: $(1\ 2)(3)$, $(1\ 3)(2)$, and $(2\ 3)(1)$. Therefore, $s(3,2) = (-1)^{3+2} \cdot 3 = -3$.

Table 6.1 Stirling numbers of the second kind $S(x,y)$ and Bell numbers $B(x)$.

x	y 1	2	3	4	5	6	7	8	9	10	$b(x)$
1	1										1
2	1	1									2
3	1	3	1								5
4	1	7	6	1							15
5	1	15	25	10	1						52
6	1	31	90	65	15	1					203
7	1	63	301	350	140	21	1				877
8	1	127	966	1,701	1,050	266	28	1			4,140
9	1	225	3,025	7,770	6,951	2,646	462	36	1		42,257
10	1	511	9,330	34,105	42,525	22,827	5,880	750	45	1	115,975

Explicit formulas for $|s(n,k)|$ are quite complicated, so we prefer a recurrence formula. Observe that $|s(n,n)| = 1$ and $|s(n,1)| = (n-1)!$ (there are $(n-1)!$ ways to seat n guests at a circular table). Considering S_n acting on the set $\{1, \ldots, n\}$, the element 1 can constitute a cycle by itself or 1 can follow one of the other $n-1$ elements in one of k cycles. In the first case, there are $|s(n-1, k-1)|$ choices for dividing the other $n-1$ elements into $k-1$ cycles. In the second case, there are $n-1$ choices for which element 1 follows, and $|s(n-1, k)|$ ways to divide $n-1$ elements into k cycles. Therefore,

$$|s(n,n)| = 1, \quad |s(n,1)| = (n-1)! \qquad (6.6)$$
$$|s(n,k)| = |s(n-1, k-1)| + (n-1)|s(n-1, k)|. \qquad (6.7)$$

From (6.6) and (6.7) we obtain a table of values of the signed Stirling numbers $s(n,k)$ for small n and k. Remember that to obtain the signed Stirling numbers we use the formula $s(n,k) = (-1)^{n+k}|s(n,k)|$.

Note that the sum of the unsigned entries of the nth row of Table 6.2 is $n!$, which is correct because each element of S_n is accounted for.

In the next sections we develop the idea of recurrence relations and generating functions, and we use generating functions to explore the connection between Stirling numbers of the first and second kind.

6.3. LINEAR RECURRENCE RELATIONS

The Fibonacci numbers $\{F_0, F_1, F_2, \ldots\}$ are defined recursively by the initial conditions $F_0 = F_1 = 1$ and the recurrence relation $F_n = F_{n-1} + F_{n-2}$ for $n \geq 2$.

Table 6.2 Stirling numbers of the first kind $s(n,k)$.

n \ k	1	2	3	4	5	6	7	8	9	10
1	1									
2	−1	1								
3	2	−3	1							
4	−6	11	−6	1						
5	24	−50	35	−10	1					
6	−120	274	−225	85	−15	1				
7	720	−1,764	1,624	−735	175	−21	1			
8	−5,040	13,068	−13,132	6,769	−1,960	322	−28	1		
9	40,320	−109,584	118,124	−67,284	22,449	−4,536	546	−36	1	
10	−362,880	1,026,576	−1,172,700	723,680	−269,325	63,273	−9,450	870	−45	1

The Fibonacci numbers occur in diverse settings, such as the following:

- F_n is the number of ways an $n \times 1$ box may be packed with 2×1 and 1×1 boxes.
- F_{n+1} is the number of binary strings of length n which do not contain the substring 00.
- F_{n+1} is the number of subsets of N_n which contain no consecutive integers.

It is easy to use the recurrence relation to compute the first terms of this sequence:

$$\{1, 1, 2, 3, 5, 8, 13, 21, 34, 55, 89, 144, 233, 377, 610, 987, 1597, \ldots\}.$$

How fast do the Fibonacci numbers grow? One might guess that they grow exponentially and, in fact, they do. In order to find the exact order of growth we first find an explicit formula for the nth term of the sequence. Assuming that $a_n = x^n$ is the general term of a sequence which satisfies the Fibonacci recurrence relation (but not necessarily the same initial conditions), then

$$x^{n+2} = x^{n+1} + x^n.$$

This equation is called the *characteristic equation* for the sequence. Assuming $x \neq 0$, we divide through by x^n and use the quadratic formula to find the two roots of the characteristic equation:

$$x_{1,2} = \frac{1 \pm \sqrt{5}}{2}.$$

We note that $x_1 \doteq 1.6$ and $x_2 \doteq -.6$.

Let $F_n = \alpha x_1^n + \beta x_2^n$ be a linear combination of these solutions, with $\alpha, \beta \in \mathcal{R}$. Then F_n is also a solution of the recurrence relation. For

$$\begin{aligned}
F_{n-1} + F_{n-2} &= \alpha x_1^{n-1} + \beta x_2^{n-1} + \alpha x_1^{n-2} + \beta x_2^{n-2} \\
&= \alpha(x_1^{n-1} + x_1^{n-2}) + \beta(x_2^{n-1} + x_2^{n-2}) \\
&= \alpha x_1^n + \beta x_2^n \\
&= F_n.
\end{aligned}$$

Thus the set of solutions is a vector space of functions of the form:

$$F_n = \alpha x_1^n + \beta x_2^n,$$

where $\alpha, \beta \in \mathcal{R}$.

We use the initial conditions to solve for the coefficients α and β. Recalling that $F_0 = 1$ and $F_1 = 1$, we obtain two linear equations to solve simultaneously:

$$1 = \alpha + \beta$$

and

$$1 = \alpha \left(\frac{1 + \sqrt{5}}{2} \right) + \beta \left(\frac{1 - \sqrt{5}}{2} \right).$$

Thus the general formula for the Fibonacci numbers is

$$F_n = \frac{5 + \sqrt{5}}{10} \left(\frac{1 + \sqrt{5}}{2} \right)^n + \frac{5 - \sqrt{5}}{10} \left(\frac{1 - \sqrt{5}}{2} \right)^n.$$

Now, how do we evaluate the growth rate of F_n? We say that a positive-valued function $f(n)$ is "big Oh" of a function $g(n)$ if

$$f(n) \le cg(n)$$

for some constant c and all sufficiently large positive integers n. We write

$$f(n) = O(g(n)).$$

If c is any constant larger than $(5 + \sqrt{5})/10$, then $F_n \le c[(1 + \sqrt{5})/2]^n$ for sufficiently large n. Thus

$$F_n = O\left(\left(\frac{1 + \sqrt{5}}{2} \right)^n \right).$$

A sequence $\{a_n\}$ is said to satisfy a *linear homogeneous recurrence relation with constant coefficients* if

$$a_n = \sum_{i=1}^{d} c_i a_{n-i},$$

for $n \ge d$.

The Fibonacci sequence satisfies a linear homogeneous recurrence relation with constant coefficients. The above procedure (with a few important modifications; see Stanley (1986)) can be used to find the general term of any such sequence.

6.4. GENERATING FUNCTIONS

With any sequence $a_0, a_1, a_2, \ldots$, finite or infinite, we associate the *ordinary generating function*

$$p(x) = \sum_{n=0}^{\infty} a_n x^n$$

and the *exponential generating function*

$$q(x) = \sum_{n=0}^{\infty} \frac{a_n x^n}{n!}.$$

The generating functions $p(x)$ and $q(x)$ contain all possible information about the sequence $\{a_n\}$, and, being algebraic entities, they are often easier to manipulate than the sequence itself. The term a_n is recovered by finding the coefficient of x^n in $p(x)$, or by finding the coefficient of x^n in $q(x)$ and multiplying by $n!$. Also, from Taylor's formula, $a_n = p^{(n)}(0)/n!$ and $a_n = q^{(n)}(0)$.

For instance, the *Fibonacci sequence* has the ordinary generating function $p(x) = \sum_{n=0}^{\infty} F_n x^n$. Exploiting the recurrence relation for F we find that $p(x) = xp(x) + x^2 p(x)$, so that $p(x) = (1 - x - x^2)^{-1}$. The function $p(x)$ contains complete information about the Fibonacci numbers and can be used to evaluate related infinite sums such as $\sum_{n=1}^{\infty} F_n(n/2^n)$. The computation of this sum is called for in the exercises.

The Fibonacci sequence illustrates the fact that a sequence satisfies a linear homogeneous recurrence relation with constant coefficients if and only if it has a rational ordinary generating function of a certain type.

Theorem 6.3. *A sequence $\{a_n\}$ satisfies a linear homogeneous recurrence relation with constant coefficients $c_1, \ldots, c_d$, i.e.,*

$$a_n = \sum_{i=1}^{d} c_i a_{n-i}$$

(for $n \geq d$), if and only if $\{a_n\}$ has a rational ordinary generating function of the form

$$\frac{f(x)}{(1 - \sum_{i=1}^{d} c_i x^i)},$$

where f is a polynomial of degree at most $d - 1$.

Proof. Suppose $a_n = \sum_{i=1}^{d} c_i a_{n-i}$ $(n \geq d)$. Then

$$p(x) = \sum_{n=0}^{\infty} a_n x^n$$

$$= \sum_{n=0}^{d-1} a_n x^n + \sum_{n=d}^{\infty} a_n x^n$$

$$= \sum_{n=0}^{d-1} a_n x^n + \sum_{n=d}^{\infty} \sum_{i=1}^{d} c_i a_{n-i} x^n$$

$$= \sum_{n=0}^{d-1} a_n x^n + \sum_{i=1}^{d} c_i \sum_{n=d}^{\infty} a_{n-i} x^n$$

$$= \sum_{n=0}^{d-1} a_n x^n + \sum_{i=1}^{d} c_i x^i \sum_{n=d-i}^{\infty} a_n x^n$$

$$= \sum_{n=0}^{d-1} a_n x^n + \sum_{i=1}^{d} c_i x^i [p(x) - \sum_{n=0}^{d-i-1} a_n x^n],$$

so that $p(x) = f(x)/(1 - \sum_{i=1}^{d} c_i x^i)$, where deg $f \leq d - 1$; and these steps are reversible. $\square$

Unfortunately, not all sequences satisfy linear recurrence relations. We now proceed to examine two nonlinear recurrence relations.

As in Section 6.1, we let D_n denote the number of derangements in S_n. We claim that D_n satisfies the linear recurrence relation

$$D_{n+1} = n(D_n + D_{n-1}).$$

For the element $n + 1$ must occur in a two-cycle or a cycle of greater length in a derangement in S_{n+1}. There are n choices for the other element in a two-cycle, and the remaining elements constitute a derangement in S_{n-1}. In a cycle of length greater than two there are n choices for the element which precedes $n + 1$, and the elements other than $n + 1$ constitute a derangement in S_n. Therefore, $D_{n+1} = n(D_n + D_{n-1})$.

Let $D(x)$ be the exponential generating function for the D_n:

$$D(x) = \sum_{n=0}^{\infty} D_n \frac{x^n}{n!}.$$

For convenience, we set $D_0 = 1$. From the recurrence relation, it follows that

$$(1 - x)D'(x) - xD(x), \tag{6.8}$$

for

$$\begin{aligned}
(1 - x)D'(x) &= \sum_{n=1}^{\infty} D_n \frac{x^{n-1}}{(n-1)!} - \sum_{n=1}^{\infty} D_n \frac{x^n}{(n-1)!} \\
&= \sum_{n=0}^{\infty} D_{n+1} \frac{x^n}{n!} - \sum_{n=1}^{\infty} D_n \frac{x^n}{(n-1)!} \\
&= \sum_{n=0}^{\infty} \frac{D_{n+1} - nD_n}{n!} x^n \\
&= \sum_{n=0}^{\infty} \frac{nD_{n-1}}{n!} x^n \\
&= \sum_{n=1}^{\infty} \frac{D_{n-1}}{(n-1)!} x^n \\
&= xD(x).
\end{aligned}$$

Separating variables in (6.8), we obtain

$$\int \frac{D'(x)dx}{D(x)} = \int \frac{xdx}{1-x},$$

from which it follows that

$$D(x) = C \frac{e^{-x}}{1-x}.$$

The condition $D_0 = 1$ implies that $C = 1$. As we mentioned earlier, the coefficients of the generating function give us a formula for the general term of the underlying sequence. In this case, we obtain the explicit formula derived in Section 6.1:

$$D_n = \sum_{j=0}^{n} (-1)^j \frac{n!}{j!}.$$

We now consider a famous sequence of numbers called *Catalan numbers*. Following the mathematician E. Catalan (1814–1894), we consider a set S together with a nonassociative operation. Suppose we have a product $a_1 \cdots a_n$ with n terms. In how many ways may the product be parenthesized? For example, with $n = 3$ there are two valid parenthesizations: $(a_1 a_2)a_3$ and $a_1(a_2 a_3)$. For $n \geq 2$, let $f(n)$ be the number of parenthesizations. A little work shows that $f(2) = 1$, $f(3) = 2$, and $f(4) = 5$. For convenience we let $f(1) = 1$. Let $f(x) = \sum_{n=1}^{\infty} f(n)x^n$ be the ordinary generating function for the Catalan numbers. The strategy is to find an equation satisified by $f(x)$, solve the equation, and read off the coefficients. Each product is composed of two parenthesized subproducts, one of m terms and the other of $n - m$ terms, where $1 \leq m \leq n - 1$. It follows that the Catalan numbers satisfy the recurrence relation

$$f(n) = \sum_{m=1}^{n-1} f(m)f(n - m).$$

It is easy to show that this recurrence relation implies

$$f(x) = x + f(x)^2.$$

From the quadratic formula, and the fact that $f(0) = 0$, it follows that

$$f(x) = \frac{1 - \sqrt{1 - 4x}}{2}.$$

From the binomial series for $\sqrt{1 - 4x}$ we obtain

$$f(n) = \frac{1}{n}\binom{2n - 2}{n - 1}.$$

The Catalan numbers occur in many settings, such as the following:

- $f(n+1)$ is the number of bit strings of length $2n$ containing exactly n 1's and n 0's, such that, reading them left to right, there are always at least as many 1's counted as 0's.
- $f(n-1)$ is the number of triangulations of a convex n-gon with $n-3$ nonintersecting diagonals.
- $f(n+1)$ is the number of binary search trees on n vertices.
- $f(n+1)$ is the number of lattice paths in the first quadrant of the plane which start at $(0,0)$, end at $(2n,0)$, and proceed at each step by $(\Delta x, \Delta y) = (+1,+1)$ or $(+1,-1)$.

6.5. SPECIAL GENERATING FUNCTIONS

In this section we determine and work with generating functions for the Stirling numbers of the first and second kind and for the Bell numbers. In Section 6.6 we determine generating functions for the partition numbers $p(n)$ and $p(n,k)$.

Define x *lower* n to be the polynomial

$$[x]_n = x(x-1)(x-2) \cdots (x-n+1),$$

and x *upper* n to be the polynomial

$$[x]^n = x(x+1)(x+2) \cdots (x+n-1).$$

There is an immediate equation connecting $[x]_n$ and $[x]^n$, namely,

$$
\begin{aligned}
[-x]_n &= -x(-x-1) \cdots (-x-n+1) \\
&= (-1)^n x(x+1) \cdots (x+n-1) \\
&= (-1)^n [x]^n.
\end{aligned}
$$

The following theorem states that $[x]^n$ is the ordinary generating function for the unsigned Stirling numbers of the first kind $|s(n,k)|$.

Theorem 6.4.

$$[x]^n = \sum_{k=1}^{n} |s(n,k)| x^k.$$

Proof. The proof is by induction on n. The case $n=1$ is trivial: $[x]^1 = x = |s(1,1)| x^1$. Assuming the result for n, it follows that

$$[x]^{n+1} = [x]^n(x+n)$$

$$= \sum_{k=1}^{n} |s(n,k)|x^k(x+n)$$

$$= \sum_{k=1}^{n+1} (|s(n,k-1)| + n|s(n,k)|)x^k$$

$$= \sum_{k=1}^{n+1} |s(n+1,k)|x^k$$

which is the correct formula for the $n+1$ case. $\square$

Application. What is the expected number of cycles in a randomly chosen permutation of S_n? Differentiating the generating function for $|s(n,k)|$ with respect to x, we obtain

$$D_x x(x+1)\ldots(x+n-1) = D_x[x]^n$$

$$= D_x \sum_{k=1}^{n} |s(n,k)|x^k.$$

$$= \sum_{k=1}^{n} |s(n,k)|kx^{k-1}.$$

Evaluating this equation at $x = 1$, we obtain

$$n!(1 + \frac{1}{2} + \frac{1}{3} + \cdots + \frac{1}{n}) = \sum_{k=1}^{n} |s(n,k)|k.$$

Dividing by $n!$, we see that the expected number of cycles is

$$1 + \frac{1}{2} + \frac{1}{3} + \cdots + \frac{1}{n},$$

an expression asymptotic to $\ln n$. For example, in a permutation on 1000 elements, we expect to find about 7 cycles.

Recalling the formula $s(n,k) = (-1)^{n+k}|s(n,k)|$, we find that $[x]_n$ is the generating function for the unsigned Stirling numbers of the first kind.

Theorem 6.5.

$$[x]_n = \sum_{k=1}^{n} s(n,k)x^k. \tag{6.9}$$

Proof.

$$[x]_n = (-1)^n[-x]^n$$

$$= (-1)^n \sum_{k=1}^{n} |s(n,k)|(-x)^k \qquad \text{(Theorem 6.4)}$$

$$= \sum_{k=1}^{n} (-1)^{n+k} |s(n,k)| x^k$$

$$= \sum_{k=1}^{n} s(n,k)x^k.$$

To test this generating function, we compute $[x]_3 = x(x-1)(x-2) = x^3 - 3x^2 + 2x$ and note that the coefficients 1, −3, 2 constitute the third row of Table 6.2 (in reverse order).

The Stirling numbers of the second kind $S(n,k)$ satisfy a relation similar to the generating function relation of Theorem 6.5.

Theorem 6.6.

$$x^n = \sum_{k=1}^{n} S(n,k)[x]_k. \tag{6.10}$$

Proof. Recalling equation (5.1), we obtain

$$x^n = \sum_{k=1}^{n} T(n,k)\binom{x}{k}$$

$$= \sum_{k=1}^{n} \frac{T(n,k)}{k!}[x]_k$$

$$= \sum_{k=1}^{n} S(n,k)[x]_k.$$

The symmetry between equations (6.9) and (6.10) is no coincidence. The vector space of polynomials with real coefficients and no constant term has as bases the two sets $B_1 = \{x^n : n \geq 1\}$ and $B_2 = \{[x]_n : n \geq 1\}$. Equation (6.9) says that $S_1 = [s(n,k)]$ is the change of basis matrix from B_2 to B_1, while equation (6.10) says that $S_2 = [S(n,k)]$ is the change of basis matrix from B_1 to B_2. Thus the two matrices S_1 and S_2 are inverses, so that $S_1 S_2 = S_2 S_1 = I$, where I is the infinite-dimensional identity matrix. In summation form, this says

$$\sum_{k=1}^{n} S(n,k)s(k,j) = \delta(n,j) \tag{6.11}$$

$$\sum_{k=1}^{n} s(n,k)S(k,j) = \delta(n,j), \tag{6.12}$$

which leads to the following wonderful inversion formula.

Theorem 6.7. *(Inversion formula for Stirling numbers). For any two real-valued functions f and g,*

$$g(n) = \sum_{k=1}^{n} S(n,k)f(k)$$

if and only if

$$f(n) = \sum_{k=1}^{n} s(n,k)g(k).$$

Proof. Assume $g(n) = \sum_{k=1}^{n} S(n,k)f(k)$. Then

$$
\begin{aligned}
\sum_{k=1}^{n} s(n,k)g(k) &= \sum_{k=1}^{n}\sum_{j=1}^{k} s(n,k)S(k,j)f(j) \\
&= \sum_{j=1}^{n}\sum_{k=j}^{n} s(n,k)S(k,j)f(j) \\
&= \sum_{j=1}^{n} \delta(n,j)f(j) \\
&= f(n).
\end{aligned}
$$

The reverse implication is proved similarly. $\square$

Application. Let t_n be the number of transitive and reflexive relations on an arbitrary n-set X. Let p_n be the number of partial orders on X. The reader may wish to verify that $p_1 = 1, p_2 = 3, p_3 = 19$ and $t_1 = 1$, $t_2 = 4, t_3 = 29$. Although there are no known formulas for t_n and p_n, the two functions are related by the inversion formula of Theorem 6.7.

Suppose $X = \{1, \ldots, n\}$ and R is a transitive and reflexive relation on X. We define a new relation R' on X as follows: $(a, b) \in R'$ if and only if $(a, b) \in R$ and $(b, a) \in R$. We claim that R' is an equivalence relation on X. Certainly R' is reflexive, as $(a, a) \in R$ implies $(a, a) \in R'$. If $(a, b) \in R'$, then $(b, a) \in R'$ (by definition), so R' is symmetric. If $(a, b) \in R$ and $(b, c) \in R'$, then $(a, b), (b, c), (b, a), (c, b) \in R$, which implies that $(a, c), (c, a) \in R$ (because R is transitive); hence, $(a, c) \in R'$, so R' is transitive.

Therefore, R' is an equivalence relation with equivalence classes $[a] = \{b \in X : (a, b) \in R \text{ and } (b, a) \in R\}$. This means that in order to construct a transitive, reflexive relation on X, we must first partition X into equivalence classes. As X may be partitioned into k equivalence classes in $S(n, k)$ ways, the question is, how are the equivalence classes pieced together? Suppose $[a]$ and $[b]$ are two different equivalence classes under R' and $(a, b) \in R$. By transitivity, $(a, c) \in R$ for all $c \in [b]$. Again, by transitivity, $(d, c) \in R$ for all $d \in [a]$. To paraphrase: "Everything in $[a]$ is arrowed to everything in $[b]$." Therefore we can think of the equivalence classes as points that are either joined or not joined by an arrow. This defines a partial order on the set of equivalence classes. In other words, the k equivalence classes are partially ordered. Because this may be done in p_k ways, summing over all possible values of k, we obtain

$$t_n = \sum_{k=1}^{n} S(n, k) p_k. \tag{6.13}$$

We test equation (6.13) by putting in the values $p_1 = 1, p_2 = 3, p_3 = 19$, $S(3, 1) = 1$, $S(3, 2) = 3$, $S(3, 3) = 1$, and calculating $t_3 = 29$.

Equation (6.13) would provide a formula for t_n if only a formula for p_n were known (as we already have formula (6.3) for $S(n, k)$). However, using the inversion formula for Stirling numbers we can write p_n in terms of t_n:

$$p_n = \sum_{k=1}^{n} s(n, k) t_k. \tag{6.14}$$

For example, plugging in the values $t_1 = 1$, $t_2 = 4$, $t_3 = 29$, $s(3,1) = 2$, $s(3,2) = -3$, $s(3,3) = 1$, we obtain $p_3 = 19$.

Open Problem 12. *Find explicit formulas for p_n and t_n.*

There are many sequences for which there is no known explicit formula for the general term. The sequences $\{p_n\}$ and $\{t_n\}$ above are well-known examples. We let p_n^* and t_n^* be the unlabeled set versions of p_n and t_n. There are no known formulas for these numbers. It can be shown that $t_n = T_n$, where T_n is the number of topologies on an n-set. Also, we define t_n' to be the number of transitive, reflexive orders on $\{1, \ldots, n\}$ which are consistent with the usual $\leq$ order on $\{1, \ldots, n\}$. There is no known formula for these numbers, although it is known that $\log_2 t_n' \sim \log_2 p_n = (n^2/4) + o(n^2)$. In Table 6.3 we present some values of these five functions.

Open Problem 13. *Find formulas for any of these functions.*

We close this section by determining the exponential generating function for the Bell numbers. From equation (6.3) we obtain an explicit formula for the nth Bell number:

$$B(n) = \sum_{k=1}^{n} \frac{1}{k!} \sum_{j=0}^{k} (-1)^{k-j} \binom{k}{j} j^n. \qquad (6.15)$$

Equation (6.15) can be simplified considerably.

Theorem 6.8.

$$B(n) = \frac{1}{e} \sum_{j=0}^{\infty} \frac{j^n}{j!}. \qquad (6.16)$$

Table 6.3 Some values of five important functions.

n	1	2	3	4	5	6	7	8	9
p_n	1	3	19	219	4,231	130,023	6,129,659	431,723,379	44,511,042,511
p_n^*	1	2	5	16	318	2 045			
t_n	1	4	29	355	6,942	209,527	9,535,241	642,779,354	63,260,289,423
t_n^*	1	3	9	32					
t_n'	1	2	7	40	357	4,824	96,428		

Proof.

$$B(n) = \sum_{k=1}^{n} S(n,k)$$

$$= \sum_{k=1}^{\infty} S(n,k)$$

$$= \sum_{k=1}^{\infty} \frac{1}{k!} \sum_{j=0}^{k} (-1)^{k-j} \binom{k}{j} j^n$$

$$= \sum_{j=0}^{\infty} \frac{j^n}{j!} \sum_{k=j}^{\infty} \frac{(-1)^{k-j}}{(k-j)!}$$

$$= \frac{1}{e} \sum_{j=0}^{\infty} \frac{j^n}{j!}.$$

Equation (6.16) is interesting from a number-theoretic point of view, as it is not at all clear *a priori* that $(1/e) \sum_{j=0}^{\infty} (j^n/j!)$ is an integer.

Now we calculate the exponential generating function for the Bell numbers $B(n)$. For convenience, we let $B(0) = 1$.

Theorem 6.9. *Let* $\beta(x) = \sum_{n=0}^{\infty} \frac{B(n)x^n}{n!}$. *Then* $\beta(x) = e^{e^x - 1}$.

Proof.

$$\beta(x) = \sum_{n=0}^{\infty} \frac{1}{n!} \sum_{j=0}^{\infty} \frac{j^n}{j!e} x^n \quad \text{(from (6.16))}$$

$$= e^{-1} \sum_{j=0}^{\infty} \frac{1}{j!} \sum_{n=0}^{\infty} \frac{(jx)^n}{n!}$$

$$= e^{-1} \sum_{j=0}^{\infty} \frac{e^{jx}}{j!}$$

$$= e^{-1} e^{e^x}$$

$$= e^{e^x - 1}.$$

We leave it to the reader to compute the first four terms of the generating function $\beta(x)$ and compare them to the known values $B(0) = 1$, $B(1) = 1$, $B(2) = 2$, $B(3) = 5$.

6.6. PARTITION NUMBERS

We have defined the partition number $p(n, k)$ to be the number of onto functions $f : X \longrightarrow Y$, $|X| = n$, $|Y| = k$, where X and Y are unlabeled sets. Also, the partition number $p(n)$ has been defined as $p(n) = \sum_{k=1}^{n} p(n, k)$. In this section we develop ordinary generating functions for these two types of partition numbers.

Suppose X and Y are unlabeled and $f : X \longrightarrow Y$ is an onto function ($|X| = n$, $|Y| = k$). For each $y \in Y$, let λ_i be the cardinality of the inverse image of y, i.e.,

$$\lambda_i = |f^{-1}\{y\}|, \qquad 1 \le i \le k.$$

Because the inverse images account for all elements of X, it follows that

$$\lambda_1 + \lambda_2 + \cdots + \lambda_k = n. \tag{6.17}$$

Furthermore, we assume the λ_i are ordered from largest to smallest:

$$\lambda_1 \ge \lambda_2 \ge \cdots \ge \lambda_k > 0. \tag{6.18}$$

Going in the other direction, it is easy to construct a unique onto function $f : X \longrightarrow Y$ such that the λ_i satisfy (6.17) and (6.18). That is to say, the functions in question are equivalent to summations which meet conditions (6.17) and (6.18). Such summations are called *partitions* of n (*into k parts*), and there are $p(n, k)$ of them. When the number of parts is unspecified, they are called *partitions* of n, and there are $p(n)$ of them.

For example,

$$p(4, 1) = 1 \quad (4 = 4)$$
$$p(4, 2) = 2 \quad (2 + 2 = 4,\ 3 + 1 = 4)$$
$$p(4, 3) = 1 \quad (2 + 1 + 1 = 4)$$
$$p(4, 4) = 1 \quad (1 + 1 + 1 + 1 = 4)$$

and

$$p(4) = p(4, 1) + p(4, 2) + p(4, 3) + p(4, 4)$$
$$= 1 + 2 + 1 + 1$$
$$= 5.$$

Tables 6.4 and 6.5 list the values of $p(n)$ and $p(n, k)$ for small n and k.

Table 6.4 Partition numbers $p(n)$ for $1 \le n \le 100$.

n	$p(n)$	n	$p(n)$	n	$p(n)$	n	$p(n)$
1	1	26	2,436	51	239,943	76	9,289,091
2	2	27	3,010	52	281,589	77	10,619,863
3	3	28	3,718	53	329,931	78	12,132,164
4	5	29	4,565	54	386,155	79	13,848,650
5	7	30	5,604	55	541,276	80	15,796,476
6	11	31	6,842	56	526,823	81	18,004,327
7	15	32	8,349	57	614,154	82	20,506,255
8	22	33	10,143	58	715,220	83	23,338,469
9	30	34	12,310	59	831,820	84	26,543,660
10	42	35	14,883	60	966,567	85	30,167,357
11	56	36	17,977	61	1,121,505	86	34,262,962
12	77	37	21,637	62	1,300,156	87	38,887,673
13	101	38	26,015	63	1,505,499	88	44,108,109
14	135	39	31,185	64	1,741,630	89	49,995,925
15	176	40	37,338	65	2,012,558	90	56,634,173
16	231	41	44,583	66	2,323,520	91	64,112,359
17	297	42	53,174	67	2,679,689	92	72,533,807
18	385	43	63,261	68	3,087,735	93	82,010,177
19	490	44	75,175	69	3,554,345	94	92,669,720
20	627	45	89,134	70	4,087,968	95	104,651,419
21	792	46	105,558	71	4,697,205	96	118,114,304
22	1,002	47	124,754	72	5,392,783	97	133,230,930
23	1,255	48	147,273	73	6,185,689	98	150,198,135
24	1,575	49	173,525	74	7,089,500	99	169,229,875
25	1,958	50	204,226	75	8,118,264	100	190,569,292

Table 6.5 Partition numbers $p(n,k)$ for $1 \le k \le n \le 10$.

n	k 1	2	3	4	5	6	7	8	9	10
1	1									
2	1	1								
3	1	1	1							
4	1	2	1	1						
5	1	2	2	1	1					
6	1	3	3	2	1	1				
7	1	3	4	3	2	1	1			
8	1	4	5	5	3	2	1	1		
9	1	4	7	6	5	3	2	1	1	
10	1	5	8	9	7	5	3	2	1	1

Open Problem 14. *Determine whether p(n) is prime for infinitely many n.*

Open Problem 15. *Determine whether p(n) is palindromic (reads the same forward as backwards in base 10) for infinitely many n.*

A partition $\lambda_1 + \cdots + \lambda_k = n$ may be pictured with a *Ferrers diagram* consisting of k rows of dots with λ_i dots in row i, $1 \le i \le k$. The Ferrers diagram for the partition $7 + 3 + 1 + 1 = 12$ is shown in Figure 6.1.

The *transpose* of a Ferrers diagram is created by writing each row of dots as a column. For example, the partition $12 = 7 + 3 + 1 + 1$ of Figure 6.1 is transposed to create the partition $12 = 4 + 2 + 2 + 1 + 1 + 1 + 1$ of Figure 6.2.

The reader may enjoy matching each partition of 4 on the previous page with its transpose. (One partition is self-transpose.)

We now give the ordinary generating function for the partition numbers $p(n)$. For convenience, we set $p(0) = 1$.

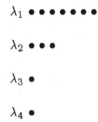

Figure 6.1. The Ferrers diagram of a partition of 12.

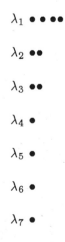

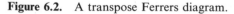

Figure 6.2. A transpose Ferrers diagram.

Theorem 6.10.

$$\sum_{n=0}^{\infty} p(n)x^n = \prod_{k=1}^{\infty}(1-x^k)^{-1}.$$

Proof. We need to show that the coefficients of x^n on the two sides of the equation are equal. The coefficient of x^n on the left side is patently $p(n)$. On the right side, the product may be written as

$$\prod_{k=1}^{\infty}(1-x^k)^{-1} = \prod_{k=1}^{\infty}(1+x^k+x^{2k}+x^{3k}+x^{4k}+\cdots).$$

To calculate the contribution to x^n from this product, suppose the term $x^{m(k)k}$ is selected from the kth factor and these terms are multiplied to yield $x^{m(1)+m(2)2+\cdots}$. If this expression equals x^n, then

$$m(1) + m(2)2 + \cdots = n. \tag{6.19}$$

Contributions to x^n correspond to solutions to (6.19). These solutions may be envisioned as Ferrers diagrams for partitions of n. With t as the greatest integer for which $m(t)$ is nonzero, we create the Ferrers diagram with $m(t)$ rows of t dots, followed by $m(t-1)$ rows of $t-1$ dots, etc. This correspondence between solutions to (6.19) and partitions of n completes the proof. $\square$

Determining the ordinary generating function for $p(n,k)$ with k fixed is a slightly more difficult matter.

Theorem 6.11.

$$\sum_{n=k}^{\infty} p(n,k)x^n = x^k \prod_{j=1}^{k}(1-x^j)^{-1}.$$

Proof. We define $p(n, \le k)$ to be the number of partitions of n into at most k parts. Considering transposes of Ferrers diagrams, $p(n, \le k)$ is also the number of partitions of n into parts of size at most k. Clearly, $p(n,k) = p(n, \le k) - p(n, \le k-1)$. Because

$$\sum_{n=k}^{\infty} p(n, \le k)x^n = \prod_{j=1}^{k}(1 + x^j + x^{2j} + x^{3j} + \cdots)$$

$$= \prod_{j=1}^{k}(1 - x^j)^{-1}$$

(by the same reasoning as that given in the proof of Theorem 6.10), we obtain

$$\sum_{n=k}^{\infty} p(n, k)x^n = \sum_{n=k}^{\infty}[p(n, \le k) - p(n, \le k - 1)]x^n$$

$$= \prod_{j=1}^{k}(1 - x^j)^{-1} - \prod_{j=1}^{k-1}(1 - x^j)^{-1}$$

$$= \prod_{j=1}^{k}(1 - x^j)^{-1}[1 - (1 - x^k)]$$

$$= x^k \prod_{j=1}^{k}(1 - x^j)^{-1}. \quad \square$$

We leave it to the reader to check the validity of the generating functions of Theorems 6.10 and 6.11 for small values of n and k.

Let $p(n, O)$ be the number of partitions of n into summands each of which is an odd number. Let $p(n, D)$ be the number of partitions of n into summands all of which are different numbers. As a further illustration of the use of generating functions, we show that $p(n, O) = p(n, D)$.

Theorem 6.12.

$$p(n, O) = p(n, D).$$

Proof.

$$\sum_{n=1}^{\infty} p(n, O)x^n = \frac{1}{(1 - x)(1 - x^3)(1 - x^5)} \cdots$$

$$= \frac{(1 - x^2)}{(1 - x)(1 - x^2)} \frac{(1 - x^4)}{(1 - x^3)(1 - x^4)} \frac{(1 - x^6)}{(1 - x^5)(1 - x^6)} \cdots$$

$$= \frac{(1-x^2)\,(1-x^4)\,(1-x^6)}{(1-x)\,(1-x^2)\,(1-x^3)} \cdots$$
$$= (1+x)(1+x^2)(1+x^3)\cdots$$
$$= \sum_{n=1}^{\infty} p(n, D)x^n.$$

The desired equality follows by comparing coefficients of the two generating functions. $\square$

Notes

The inclusion–exclusion principle was first studied by D. A. Da Silva in 1854. It was studied by J. Sylvester in 1883 and is sometimes referred to as Sylvester's cross-classification principle.

J. Sylvester introduced the notion of Ferrers diagrams in 1853. Apparently they were discovered by N. M. Ferrers.

Fibonacci (Leonardo of Pisa) introduced and discussed the Fibonacci numbers in his *Liber Abaci* in 1202.

G. H. Hardy and S. Ramanujan were the first mathematicians to find an explicit formula for $p(n)$. The most elementary asymptotic estimate is $\log_e p(n) \sim \pi(2n/3)^{1/2}$. See Hall (1986) for details.

Exercises

6.1   (Putnam Competition, 1958) Show that the number of nonzero terms in the expansion of the nth-order determinant having zeros in the main diagonal and ones elsewhere is

$$n!\left[1 - \frac{1}{1!} + \frac{1}{2!} - \frac{1}{3!} + \cdots + \frac{(-1)^n}{n!}\right].$$

[Hint: Make a connection between the terms in the expansion of this determinant and the collection of derangements of an n-element set.]

6.2   (Abel-type formulas) Find closed form expressions for $\sum_{i=0}^{\infty} ix^i$ and $\sum_{i=0}^{\infty} i^2 x^i$.

6.3   Use the technique of the previous exercise to evaluate the sum $\sum_{n=1}^{\infty} F_n n/2^n$.

6.4   (Putnam Competition, 1962) Evaluate in closed form

$$\sum_{k=1}^{n} \binom{n}{k} k^2.$$

6.5   (Series multisection) Let $\sum_{n=0}^{\infty} a_n x^n = p(x)$. Show that

$$\sum_{k=0}^{\infty} a_{qk+p} x^{qk+p} = q^{-1} \sum_{j=0}^{q-1} \omega^{-jp} p(\omega^j x),$$

where ω is a primitive qth root of unity.

6.6   Suppose an object travels along the lattice points of the plane, starting at the point $(0,0)$. At each step, the object moves one unit to the right or one unit up. The object stops when it reaches the line $y = N$ or the line $x = N$. Show that the expected length of the object's path is $2N - N\binom{2N}{N}2^{1-2N}$.

6.7   Show that

$$\sum_{k=0}^{[n/2]} \left[\binom{n}{k} - \binom{n}{k-1} \right]^2 = \frac{1}{n+1} \binom{2n}{n}.$$

6.8   Show that the Bell numbers $B(n)$ satisfy the recurrence formula $B(n+1) = \sum_{k=0}^{n} \binom{n}{k} B(k)$, where $B(0) = 1$.

6.9   Show that $\sum_{k=0}^{n} s(n,k) B(k) = 1$.

6.10   Prove that $S(n,2) = 2^{n-1} - 1$ and $S(n,n-1) = \binom{n}{2}$ for all $n \geq 2$.

6.11   Let $|X| = m$, $|Y| = n$.

(a)   How many possible relations are there from X to Y?

(b)   How many relations are there if X is unlabeled and Y is labeled?

(c)   How many relations are there if X is labeled and Y is unlabeled?

(d)   How many relations are there if both X and Y are unlabeled?

[Hint: In each case, the relation $R : X \longrightarrow Y$ may be viewed as a function $f : X \longrightarrow P(Y)$, defined by $f(x) = R(x)$. Now use the techniques associated with the fundamental counting problem for functions. For example, the answer for (b) is $\binom{m+2^n-1}{m}$.]

6.12   Prove that

$$[x+y]^n = \sum_{k=0}^{n} \binom{n}{k} [x]^k [y]^{n-k}$$

and that

$$[x+y]_n = \sum_{k=0}^{n} \binom{n}{k} [x]_k [y]_{n-k}.$$

6.13 Let a_n be the number of permutations in S_n which alternately rise, fall, rise, fall, etc. For example, 142 635 is such a permutation. Find $\sum_{n=1}^{\infty} a_n x^n$ and use this information to find a_6.

6.14 Prove the Möbius inversion formula, of which the inversion formula of Theorem 6.7 is a special case: If

$$\sum_{k=1}^{\infty} \alpha(n,k)\beta(k,j) = \sum_{k=1}^{\infty} \beta(n,k)\alpha(k,j) = \delta(n,j),$$

then $f(n) = \sum_{k=1}^{\infty} \alpha(n,k)g(k)$ if   and   only   if $g(n) = \sum_{k=1}^{\infty} \beta(n,k)f(k)$.

6.15 Let $\alpha(n,k) = 1$ if $k \mid n$ and 0 otherwise. Let $\beta(n,k)$ be the inverter for $\alpha(n,k)$ given by the Möbius inversion formula. Show that $\beta(n,k) = \mu(n/k)$, where $\mu(m) = 1$ if $m = 1$, $(-1)^k$ if m is the product of k distinct primes, and 0 otherwise.

6.16 Show that

$$\delta(n,k) = \sum_{j=0}^{n} (-1)^{j+k} \binom{n}{j} \binom{j}{k}.$$

Deduce   that $a_n = \sum_{k=0}^{n} (-1)^k \binom{n}{k} b_k$ if   and   only   if $b_n = \sum_{k=0}^{n} (-1)^k \binom{n}{k} a_k$.

6.17 We say that two sets A and B are *linked* if $A \cap B \neq \emptyset$ and neither A nor B is a subset of the other. If S is an n-element set, how many pairs (A, B) of subsets of S exist with A and B linked?

6.18 Let $|X| = m$. An *algebra* on X is a subset S of $P(X)$ with the following properties:

1. $X \in S$.
2. $A \in S$ implies $X - A \in S$.
3. $A \in S$ and $B \in S$ implies $A \cup B \in S$.

(a) How many algebras are possible on X if X is labeled?
(b) How many if X is unlabeled?

6.19 Show that $p(n,k) = \sum_{j=1}^{k} p(n-k,j)$.

6.20 Find formulas for $p(n,1)$, $p(n,2)$, and $p(n,3)$. Find an asymptotic estimate for $p(n,k)$ (with k fixed).

6.21 Show that the number of partitions of n into summands none of which occurs exactly once is the same as the number of partions of n into summands none of which is congruent to 1 or 5 modulo 6.

6.22 We say that two permutations σ and τ of S_n are in the same *conjugacy class* if there exists a permutation $\rho \in S_n$ such that $\tau = \rho\sigma\rho^{-1}$. Prove that two permutations are in the same conjugacy class if and only if they have the same cycle structure. How many conjugacy classes of S_n are there?

6.23 How many nonisomorphic abelian groups of order 2700 are there?

6.24 How many ways may one make $2.23 postage using 1 cent, 2 cent, 3 cent, 10 cent, 20 cent, $1, and $2 stamps, and not more than three of any one denomination? Notice that the answers to this exercise and the previous one are the same. Why are the answers the same?

7

Permutations and Tableaux

In this chapter we consider a problem raised in Chapter 3 concerning sequences without monotonic subsequences of given lengths. In particular, we showed that every permutation of the integers $1, \ldots, 10$ contains a monotonic subsequence (increasing or decreasing) of length four. However, the same result is not true if we consider permutations of $1, \ldots, 9$. Now we ask, how many permutations of $1, \ldots, 9$ do *not* contain a monotonic subsequence of length four?

7.1. ALGORITHM: LISTING PERMUTATIONS

Consider the following ordering of the 24 permutations of $\{1, \ldots, n\}$:

1234, 1243, 1324, 1342, 1423, 1432, 2134, 2143, 2314, 2341, 2413, 2431, 3124, 3142, 3214, 3241, 3412, 3421, 4123, 4132, 4213, 4231, 4312, 4321.

This ordering is called the *lexicographic*, or *dictionary* order. The permutations are listed in the order they would appear in a dictionary if the "alphabetical" order of the numbers were $1, 2, 3, 4$.

In order to verify claims about the set of permutations, or to generate new conjectures, it is desirable to have an algorithm for listing the permutations in some convenient order, such as the lexicographic order. There are several algorithms which do this. Perhaps the simplest procedure is to start with the permutation 1234, and then successively generate the next permutation in the lexicographic ordering, until 4321 is reached. How do we find the next permutation in the lexicographic order? As an illustration, let's consider the permutation following 3241. We should leave unaltered as many digits as possible at the left. However, the next permutation cannot be of the form 32xx because the digits which follow 32 in 3241 are already in decreasing order. Can the next permutation be of the form 3xxx? Yes, because the digits 241 are not already in decreasing order. The digit placed in the second position must be greater than 2. The only available digit is 4. Thus the permutation is of the form 34xx, and the least such permutation is 3412. In general, we must find the rightmost number a whose right neighbor b satisfies $a < b$. Then a is exchanged for the next greater available number, and the following places are filled with the available numbers in increasing order. This procedure is implemented by the following algorithm.

Algorithm: Generating permutations of n.

1. Input n.
2. Let $x_1, \ldots, x_n = 1, \ldots, n$.
3. Print $x_1, \ldots, x_n$ (the first permutation).
4. Let i^* be the greatest value of i for which $x_i < x_{i+1}$.
5. Let x_j be the smallest number for which $x_{i^*} < x_j$ and $i^* < j$.
6. Interchange the values of x_{i^*} and x_j.
7. Let $x_{i^*+1}, \ldots, x_n = x_n, \ldots, x_{i^*+1}$.
8. Print $x_1, \ldots, x_n$ (the current permutation).
9. If $x_1, \ldots, x_n = n, \ldots, 1$, then stop. Otherwise go to step 4.

The reader may wish to write a computer program based on this algorithm to enumerate the permutations of $\{1, \ldots, 9\}$ which contain no monotonic subsequence of length 4. This result is needed in Section 7.2.

7.2.   YOUNG TABLEAUX

In Section 3.1 we observed that the numbers $1, \ldots, 9$ may be arranged in a sequence with no monotonic subsequence of length 4. How many such permutations are there? A computer search of all 362,880 permutations of the numbers $1, \ldots, 9$ (see Section 7.1) reveals that there are 1764 permutations that do not have a monotonic subsequence of length 4. The alert reader may notice that $1764 = 42^2$, and therein lies the explanation. Such permutations of $1, \ldots, 9$ correspond, via the famous *Robinson–Schensted* correspondence, to ordered pairs of standardized Young tableaux of shape $3 + 3 + 3$ (fillings of a 3×3 grid with the numbers $1, \ldots, 9$ so that the numbers increase in each row and column). The so-called *hook-length* formula calculates the number of such tableaux, which in this case turns out to be 42. Therefore the number of desired sequences is $42^2 = 1764$. For a complete discussion of Young tableaux and the hook-length formula the reader is referred to Van Lint and Wilson (1992). In the present two sections we merely state some of the basic facts pertaining to Young tableaux and the Robinson–Schensted correspondence.

Let $n = \lambda_1 + \cdots + \lambda_n$. A *Young tableau* of *shape* $\lambda_1 + \cdots + \lambda_n$ is a Ferrers diagram of shape $(\lambda_1, \ldots, \lambda_n)$ in which each dot has been replaced by a different integer from among the integers $1, \ldots, n$. If the integers increase in every column and in every row, then we say the tableau is *standardized*. The positions in a tableau are called *cells*. The number n is the *order* of the tableau.

Figure 7.1 shows an example of a standardized Young tableau with $n = 9 = 4 + 3 + 2$.

How many standardized Young tableaux have this shape? The answer is given by the *hook-length formula*. We define the *hook-length* of a cell in a tableau to be one more than the number of cells to the right and below the given cell. Figure 7.2 shows the hook-lengths of each cell of the tableau of Figure 7.1.

The hook-length formula says that the number of standardized Young tableau of this shape is $n!$ divided by the product of the hook-lengths. Therefore there are

$$\frac{9!}{6 \cdot 5 \cdot 4 \cdot 3 \cdot 3 \cdot 2} = 168$$

different standardized Young tableaux of shape $4 + 3 + 2$.

We know from Section 6.6 that $p(9) = 30$. That is, there are 30 partitions of the number 9. For each partition λ of 9, let f_λ be the number of standardized tableaux given by the hook-length fomula. The reader who computes the numbers f_λ for each λ may be surprised to find that

$$\sum_\lambda f_\lambda^2 = 9!.$$

1 3 4 8

2 5 9

6 7

Figure 7.1. A standardized Young tableau of shape $4 + 3 + 2$.

Figure 7.2. The hook-lengths of the cells of a tableau.

Generalized to any value of n, this identity is known as Schur's formula. It is used in the theory of representations of the symmetric group.

Theorem 7.1. *(Schur's formula). Let $n = \lambda_1 + \cdots + \lambda_n$ and let f_λ be the number of standardized Young tableaux of shape $\lambda_1 + \cdots + \lambda_n$. Then*

$$\sum_\lambda f_\lambda^2 = n!.$$

Schur's theorem indicates that there is a correspondence between pairs of Young tableaux of identical shape and permutations of n. The aim of the next section is to show this correspondence.

7.3. THE ROBINSON–SCHENSTED CORRESPONDENCE

The Robinson–Schensted correspondence is a bijection between the group S_n of permutations on the set $\{1, \ldots, n\}$ and the set of ordered pairs (P, Q) of order n standardized Young tableaux, where P and Q have the same shape. The correspondence is effected by an algorithm, half of which (the forward half) we describe here.

Suppose we are given a permutation $\sigma \in S_n$. For example, let $n = 9$ and

$$\sigma = 583276419.$$

We read the permutation from left to right and construct P step by step. (We will show how to construct Q later.) The 5 is placed in the top left position of the tableau, and the 8, being greater than 5, is placed below:

5
8

Now we come to the 3. Being less than 5, the 3 "bumps" the 5 to the right and takes its place:

$$3\ 5$$
$$8$$

Likewise, 2 is greater than 3, so it bumps the 3 to the right and takes its place:

$$2\ 3\ 5$$
$$8$$

Now comes the 7. Because 7 is less than 8, it is inserted into the second row, bumping the 8 to the right. Then the 6 bumps the 7 (and the 8 along with it).

$$2\ 3\ 5$$
$$6\ 7\ 8$$

Now the 4 bumps the entire second row to the right, and the 8 is bumped up to the first row:

$$2\ 3\ 5\ 8$$
$$4\ 6\ 7$$

Now the 9, being greater than 4, is placed in the third row, and the 1 bumps the first row to the right:

$$1\ 2\ 3\ 5\ 8$$
$$4\ 6\ 7$$
$$9$$

This completes the tableau P. The tableau Q is constructed by putting the numbers $1, \ldots, 9$ in a tableau of the same shape as P, and in the order in which new positions were occupied in P. Figure 7.3 shows P and Q.

It so happens that the number of columns of the tableaux P and Q equals the length of the longest decreasing subsequence of the permutation. The number of rows equals the length of the longest increasing subsequence. See Stanton and White (1986). In our example, the tableaux have five columns and three rows, and, indeed, the longest

1 2 3 5 8   1 3 4 7 9

4 6 7   2 5 6

9   8

Figure 7.3. The tableaux P and Q corresponding to σ.

decreasing subsequence of σ has five terms and the longest increasing subsequence has three terms.

To calculate the number of permutations of $\{1, \ldots, n\}$ with no monotonic subsequence of length 4, we simply use the hook-length formula to find the number of standardized Young tableaux of shape $3 + 3 + 3$:

$$\frac{9!}{2 \cdot 2 \cdot 3 \cdot 3 \cdot 3 \cdot 4 \cdot 5} = 42.$$

Therefore the number of such permutations is $42^2 = 1764$.

Notes

For more algorithms concerning Young tableaux, the reader is referred to Stanton and White (1986).

Exercises

7.1 Explain how the reverse half of the Robinson–Schensted algorithm works.

7.2 Suppose the Robinson–Schensted correspondence associates the permutation σ with the pair of tableaux (P, Q). Show that the number of columns of the tableaux P and Q equals the length of the longest decreasing subsequence of σ. Show that the number of rows in P and Q equals the length of the longest increasing subsequence of σ.

7.3 Suppose the permutation σ corresponds to the pair of tableaux (P, Q). Show that σ^{-1} corresponds to the pair (P, Q).

7.4 A permutation $\sigma \in S_n$ is called an *involution* if $\sigma = \sigma^{-1}$. Show that the number of involutions in S_n equals the number of standardized tableaux of order n.

8

The Pólya Theory of Counting

In Chapter 5 we classified functions $f : X \longrightarrow Y$ ($|X| = x$, $|Y| = y$), where X and Y are labeled or unlabeled sets; and in Chapter 6 we enumerated these functions. In the present chapter the notions of labeled and unlabeled sets are generalized. For instance, recall that two functions $f : X \longrightarrow Y$ and $g : X \longrightarrow Y$ are equivalent in the X unlabeled, Y labeled sense if there exists a bijection $h : X \longrightarrow X$ such that $f = gh$. The bijection h can be viewed as a permutation of X (in fact, *any* permutation in the symmetric group S_x). What happens if we restrict the permutations to a specified subgroup G of S_x? If G is the identity group (e), for example, then we obtain the X labeled case; while if $G = S_x$, then we obtain the X unlabeled case. The nontrivial subgroups G give rise to interesting intermediate cases. In these cases, Pólya's theorem for the number of inequivalent functions allows us to count quite complicated configurations, including nonisomorphic graphs. More generally, if one group G acts on X and another group H acts on Y, then the number of inequivalent functions is given by De Bruijn's formula, which enumerates more intricate structures, such as self-complementary graphs.

8.1. BURNSIDE'S LEMMA

Let G be a finite group and X a finite nonempty set. An *action* of G on X is a function $\theta : G \times X \longrightarrow X$ which satisfies the following two conditions:

1. For every $x \in X$, $\theta(e, x) = x$ (where e is the identity element of G).
2. For every $g, h \in G$ and $x \in X$, $\theta(g, \theta(h, x)) = \theta(gh, x)$.

For convenience, we write $\theta(g, x)$ as gx, so that the two conditions become

117

1. $ex = x$.
2. $g(hx) = (gh)x$.

Remember, however, that g and h are group elements and x is a set element.

Example. The symmetric group S_n acts on $\mathcal{N}_n$ by the *natural action* $gx = g(x)$, where $g(x)$ is the image of x under the function $g : \mathcal{N}_n \longrightarrow \mathcal{N}_n$.

Example. The cyclic group $\mathcal{Z}_n$ acts on $\mathcal{N}_n$ by the action

$$gx = g + x \quad \text{for} \quad g + x \leq n \tag{1}$$
$$gx = g + x - n \quad \text{for} \quad g + x > n.$$

(Here g denotes the equivalence class representative of $[g]$ that lies between 1 and n.)

If $x \in X$, then the *orbit* of x (under the action θ) is orb$(x) = \{gx : g \in G\}$. Orbits constitute equivalence classes of X; that is, X is partitioned into orbits. If there is only one orbit, then the action θ is *transitive*. Both examples above are transitive actions.

To find the size of orb(x), note that $gx = hx$ if and only if $h^{-1}gx = x$, which is true if and only if $h^{-1}g \in G_x$, where $G_x = \{g \in G : gx = x\}$. ($G_x$ is the *stabilizer* of x.) Thus $gx = hx$ if and only if $g \in hG_x$, which holds if and only if g and h are in the same coset of G_x. Therefore, the number of distinct values of gx is the number of cosets of G_x:

$$|\text{orb}(x)| = [G : G_x] = \frac{|G|}{|G_x|}.$$

Example. Suppose G is a finite group. For each $g \in G$, define $f_g : G \to G$ by $f_g(x) = gxg^{-1}$. We say that G acts on itself by *conjugation*. The stabilizer of an element $x \in G$ is called the *centralizer* of x and is denoted $C(x)$. Let ccl(x) denote the conjugacy class of x. The size of the conjugacy class of x is given by the formula

$$|\text{ccl}(x)| = |G|/|C(x)|.$$

Two permutations are in the same conjugacy class if and only if they have the same cycle structure (Exercise 6.22). Therefore, the number of conjugacy classes equals the partition number $p(n)$, where $n = |G|$.

Theorem 8.1. *(Burnside's lemma). Let G be a finite group acting on a finite nonempty set X. For each $g \in G$, let f_g be the number of elements of X fixed by g. The number of orbits n is given by*

$$n = \frac{1}{|G|} \sum_{g \in G} f_g. \tag{8.1}$$

Proof. The proof is a nice example of the technique of enumerating a set two different ways and comparing the results. In this case, the set is $S = \{(g, x) : g \in G, \ x \in X, \text{ and } gx = x\}$. On one hand, by definition of f_g, $|S| = \sum_{g \in G} f_g$. On the other hand, counting from the perspective of the elements of X, and letting x' denote orbit representatives, we obtain

$$|S| = \sum_{x \in X} |G_x|$$

$$= \sum_{x \in X} \frac{|G|}{|\mathrm{orb}(x)|}$$

$$= |G| \sum_{x \in X} \frac{1}{|\mathrm{orb}(x)|}$$

$$= |G| \sum_{x'} \frac{|\mathrm{orb}(x')|}{|\mathrm{orb}(x')|}$$

$$= |G| \sum_{x'} 1$$

$$= |G|n.$$

Equating the two expressions for $|S|$, we obtain $\sum_{g \in G} f_g = |G|n$, from which (8.1) follows instantly. $\square$

Application. Now we are able to give an elegant proof of the assertion made at the beginning of Part II. (Recall the complicated proof of Section 6.1.) How many fixed points has the average permutation of S_n? To determine the average when $n = 3$, we list the permutations of S_3 and the number of fixed points of each.

Permutation	Number of fixed points
(1)(2)(3)	3
(1)(2 3)	1
(2)(1 3)	1

(3)(1 2)	1
(1 2 3)	0
(1 3 2)	0

The total number of fixed points is 6, and the average number is $\frac{6}{6} = 1$. It turns out that the average is always 1, regardless of n. Applying Burnside's lemma to the natural action of S_n on N_n (which is transitive), we obtain

$$1 = \frac{1}{n!} \sum_{g \in S_n} f_g,$$

the claimed average.

We now give a third proof of this result using the probabilistic method of Section 4.4. Randomly choose a permutation of S_n, and let X_k be a random variable equal to 1 if k is fixed by this permutation and 0 otherwise. By direct calculation, the expected value of X_k is $E(X_k) = (n-1)!/n! = 1/n$. Let X be the sum of all X_k. By linearity of expectation,

$$E(X) = \sum_{k=1}^{n} E(X_k) = n \cdot \frac{1}{n} = 1.$$

In the next section we apply Burnside's lemma to functions $f : X \longrightarrow Y$, thus generalizing the labeled/unlabeled sets of Chapters 5 and 6.

8.2. LABELINGS

In this section we assume that $f : X \longrightarrow Y$ is a function from a set X of m elements to a set Y of n elements. In the terminology of Pólya's theory of counting, the elements of Y are *labels*, and f is a *labeling* of X. Suppose that a finite group G acts on X. We picture this situation with the following diagram:

$$f : \overset{G}{\underset{\downarrow}{X}} \longrightarrow Y.$$

This action of G on X induces an action of G on the set of labelings Y^X as follows: $(gf)(x) = f(g^{-1}x)$ for all $x \in X$. We need to check that the two axioms for an action are satisfied:

1. $e(f(x)) = f(e^{-1}x) = f(x)$ for all $x \in X$.
2. $(g(hf))(x) = (hf)(g^{-1}x) = f(h^{-1}g^{-1}x)$
   $= f((gh)^{-1}x) = ((gh)f)(x)$ for all $x \in X$.

The set Y^X of functions is partitioned into equivalence classes by this action, and functions in different equivalence classes are called *G-inequivalent functions*. By definition, the number of G-inequivalent functions is the number of orbits of the action of G on Y^X.

Theorem 8.2. *When G (finite) acts on the set of functions Y^X ($|X| = m$, $|Y| = n$), the number of orbits is*

$$\frac{1}{|G|}\sum_{g \in G} n^{c(1)+c(2)+\cdots+c(m)}, \tag{8.2}$$

where $c(i)$ is the number of cycles of length i of g (regarded as a permutation of X).

Proof. Suppose g, when regarded as a permutation of X, has cycle structure $c = (c(1), c(2), \ldots, c(m))$. The functions fixed by g are precisely those that are constant on each cycle. As there are $c(1) + \cdots + c(m)$ cycles, each of which may be assigned one of n images, the number of functions fixed by g is $f_g = n^{c(1)+\cdots+c(m)}$. Our conclusion now follows directly from Burnside's lemma. $\square$

Application. How many stacks of 11 poker chips are possible with two colors of poker chips (red and white)? Let X be the set of 11 positions in the stack and $Y = \{red, white\}$. The group $S_2 = \{e, \tau\}$ acts on X, the identity e leaving the stack alone and τ turning the stack upside-down. To use Theorem 8.2, we need to know the cycle structures of e and τ. Certainly e consists of 11 fixed points, so $c(1) = 11$ and $c(i) = 0$ for $2 \le i \le 11$. And τ consists of 1 fixed point (the middle poker chip) and 5 transpositions, so $c(1) = 1$, $c(2) = 5$, and $c(i) = 0$ for $3 \le i \le 11$. Therefore, the number of S_2-inequivalent stacks is

$$\tfrac{1}{2}(2^{11} + 2^6) = 1056.$$

While it is theoretically possible at this point to enumerate some rather complicated structures (nonisomorphic graphs, for instance), we prefer to do so only after developing some additional machinery – cycle indexes – in the next section.

8.3. CYCLE INDEXES

As in the previous section, we let $c = (c(1), \ldots, c(m))$ be the cycle structure of $g \in G$ when G acts on X (G finite, $|X| = m$). To g we assign the monomial

$$x_1^{c(1)} x_2^{c(2)} \cdots x_m^{c(m)},$$

where the x_i are place-keeping variables in a commutative ring containing the rational numbers. The *cycle index* $Z(G)$ is the average of these monomials:

$$Z(G) = \frac{1}{|G|} \sum_{g \in G} x_1^{c(1)} x_2^{c(2)} \cdots x_m^{c(m)}.$$

The cycle index $Z(G)$ stores complete information about the cycle lengths of the various permutations g of the group action. G. Pólya (1887–1985) chose the letter Z to stand for the German word *Zycklen-zeiger* ("cycle indicator"). As Pólya said, "The cycle index knows many things." For instance, letting each $x_i = n$ (a substitution denoted $Z(G)[x_i \longleftarrow n]$), we obtain formula (8.2) for the number of G-inequivalent functions in Y^X ($|X| = m$, $|Y| = n$):

$$Z(G)[x_i \leftarrow n] = \frac{1}{|G|} \sum_{g \in G} n^{c(1) + \cdots + c(m)}.$$

We will find out about other substitutions in Section 8.4. For now, we calculate the cycle indexes of the most important group actions: E_n, $\mathcal{Z}_n$, D_n, A_n, S_n. The set acted upon is always $X = \{1, \ldots, n\}$.

   1. **The identity group E_n.** The identity group consists of only the
      identity element e, which fixes every element of X. There are n
      1-cycles, and the cycle index is

$$Z(E_n) = x_1^n. \tag{8.3}$$

   2. **The cyclic group $\mathcal{Z}_n$.** Given $g \in G$, $x \in X$, the length of the cycle
      containing x is the minimum positive k for which

$gk + x \equiv x \bmod n$, or $gk \equiv 0 \bmod n$. Because k is independent of x, all cycles of g have length k. If g contains j cycles, then $jk = n$, from which it follows that k is a divisor of n and $j = n/k$. To determine the number of elements g corresponding to each value of k, observe that $k'n/k$ has order k whenever $\gcd(k', k) = 1$. There are $\phi(k)$ such values of k', by definition of Euler's ϕ-function. As $\sum_{k|n} \phi(k) = n$ (Exercise 1.28), there are exactly $\phi(k)$ values of g for each k/n. Therefore,

$$Z(\mathcal{Z}_n) = \frac{1}{n} \sum_{k|n} \phi(k) x_k^{n/k}. \tag{8.4}$$

3. **The dihedral group D_n.** As D_n contains $\mathcal{Z}_n$ as a subgroup, the cycle index of D_n will contain all the terms in (8.4). The other elements of D_n are "flips". If n is odd, each flip fixes one element of X and contains $(n-1)/2$ transpositions. If n is even, half of the flips contain $n/2$ transpositions and half contain two fixed points and $(n-2)/2$ transpositions. Putting these facts together, we obtain the formulas

$$Z(D_n) = \begin{cases} \frac{1}{2} Z(\mathcal{Z}_n) + \frac{1}{2} x_1 x_2^{(n-1)/2} & (n \text{ odd}) \tag{8.5} \\ \frac{1}{2} Z(\mathcal{Z}_n) + \frac{1}{4} (x_2^{n/2} + x_1^2 x_2^{(n-2)/2}) & (n \text{ even}). \tag{8.6} \end{cases}$$

4. **The symmetric group S_n.** It is convenient to discuss the symmetric group before the alternating group. A permutation $g \in S_n$ can have any cycle structure $c = (c(1), \ldots, c(n))$, where

$$1c(1) + 2c(2) + \cdots + nc(n) = n, \tag{8.7}$$

as each side of this equation counts 1 for every element of X. The number of solutions to (8.7) is a good counting puzzle in its own right. Solutions to (8.7) may be generated by ordering the elements of X and partitioning the elements from left to right into cycles of the appropriate lengths. Each of the $n!$ orderings of X gives rise to repeated solutions due to interchanging cycles and writing cycles down in more than one way. Because there are $c(k)!$ ways to list cycles of length k, and each cycle may be written k ways, the number of solutions to (8.7) is

$$h(c) = \frac{n!}{\prod_{k=1}^{n} k^{c(k)} c(k)!}, \tag{8.8}$$

and we obtain

$$Z(S_n) = \frac{1}{n!} \sum_c h(c) x_1^{c(1)} \cdots x_n^{c(n)}. \tag{8.9}$$

5. The alternating group A_n. The alternating group consists of the $\frac{1}{2}n!$ even permutations of S_n. When a permutation is written as a disjoint product of cycles, it is easy to tell whether it is even or odd. Because each odd cycle is equal to the product of an even number of transpositions, the number of odd cycles has no bearing on whether a permutation is even. However, each even cycle is equal to the product of an odd number of cycles. Therefore, in order for a permutation to be even, it must be composed of an even number of disjoint cycles of even length. (This is the same logic as used in the handshake theorem.) Therefore,

$$\frac{1}{2}(1 + (-1)^{c(2)+c(4)+\cdots})$$

counts 1 for every even permutation and 0 for every odd permuation in S_n. This establishes the following cycle index:

$$Z(A_n) = \frac{1}{n!} \sum_c h(c)(1 + (-1)^{c(2)+c(4)+\cdots}) x_1^{c(1)} \cdots x_n^{c(n)}. \tag{8.10}$$

Application. How many 11-bead necklaces can be made with two types of beads? The appropriate group is D_{11} acting on the set $X = \{1, \ldots, 11\}$. By formulas (8.4) and (8.5),

$$Z(D_{11}) = \frac{1}{2} Z(\mathcal{Z}_{11}) + \frac{1}{2} x_1 x_2^5$$

$$= \frac{1}{22} \sum_{k|11} \phi(k) x_k^{n/k} + \frac{1}{2} x_1 x_2^5.$$

The only divisors of 11 are 1 and 11, for which $\phi(1) = 1$ and $\phi(11) = 10$. Thus,

$$Z(D_{11}) = \frac{1}{22}(x_1^{11} + 10 x_{11}) + \frac{1}{2} x_1 x_2^5.$$

The number of necklaces which can be made with two types of beads is obtained by making the substitution

$$Z(D_{11})[x_i \leftarrow 2] = \tfrac{1}{22}(2^{11} + 10 \cdot 2) + \tfrac{1}{2} \cdot 2 \cdot 2^5$$
$$= 126.$$

Recall that in Section 8.2 we determined that there are 1056 different stacks of 11 poker chips using two types of chips. The number of necklaces with 11 beads of two types is necessarily smaller because a necklace has more symmetries than a stack of chips.

8.4. PÓLYA'S THEOREM

Theorem 8.3. *(Pólya's theorem). Let G act on X, and therefore on the set of functions $f : X \longrightarrow Y$ (G finite, $|X| = m$, $|Y| = n$). Suppose $F(y(1), \ldots, y(n))$ is the set of G-inequivalent functions for which $|f^{-1}(y_i)| = y(i)$, $1 \le i \le n$. Then*

$$Z(G)[x_1 \leftarrow \sum_{j=1}^{n} y_j^i] = \sum |F(y(1), \ldots, y(n))| y_1^{y(1)} \cdots y_n^{y(n)} \qquad (8.11)$$

where the right-hand sum is taken over all n-tuples $(y(1), \ldots, y(n))$ with

$$\sum_{j=1}^{n} y(j) = m.$$

Proof.   We need to show that the coefficient of $y_1^{y(1)} \cdots y_n^{y(n)}$ on the left-hand side of (8.11) is $|F(y(1), \ldots, y(n))|$. By Burnside's lemma,

$$|F(y(1), \ldots, y(n))| = \frac{1}{|G|} \sum_{g \in G} f_g, \qquad (8.12)$$

where f_g is the number of functions in $F(y(1), \ldots, y(n))$ fixed by g.
    Suppose g has cycle structure $c = (c(1), \ldots, c(n))$. If the function

$$f \in F(y(1), \ldots, y(n))$$

is fixed by g, then f is constant on each cycle of g; that is, each cycle of g lies entirely within one of the inverse images $f^{-1}(y_i)$. Picture an $m \times 1$ box (the elements of X) partitioned into n sections of sizes $y(1), \ldots, y(n)$

(the inverse images of f). Then f_g is the number of possible packings of this box with $c(i)$ blocks of sizes i, $1 \leq i \leq m$ (the cycles of g). The polynomial on the left-hand side of (8.11) is

$$\frac{1}{|G|} \sum x_1^{c(1)} \cdots x_m^{c(m)} [x_i \longleftarrow \sum_{j=1}^{n} y_j^i] = \frac{1}{|G|} \sum_{g \in G} \prod_{k=1}^{m} (\sum_{j=1}^{n} y_j^k)^{c(k)}. \quad (8.13)$$

Let us consider how the $y_1^{y(1)} \cdots y_n^{y(n)}$ terms are formed in (8.13). Suppose each multiplicand $(\sum_{j=1}^{n} y_j^i)^{c(i)}$ is expanded as a product of $c(i)$ factors. Then each term in the product of these multiplicands is obtained by choosing a y_j^i from each factor. The contribution to $y_1^{y(1)} \cdots y_n^{y(n)}$ is the number of ways the exponents ($c(i)$ units of size i) may be arranged to equal $y(i)$, for each $1 \leq i \leq n$. These arrangements are clearly equivalent to the box packings described above. Therefore, the coefficient of $y_1^{y(1)} \cdots y_n^{y(n)}$ is

$$\frac{1}{|G|} \sum_{g \in G} f_g = |F(y(1), \ldots, y(n))|,$$

as we needed to show. $\square$

In order to use Pólya's theorem to count nonisomorphic graphs, we need to determine the cycle index of the appropriate group action. A graph $G = (V, E)$ may be identified with a function $f : [V]^2 \longrightarrow \{0, 1\}$, where $f(\{x, y\}) = 1$ or 0 according to whether $\{x, y\}$ is or is not a member of E. Two graphs are isomorphic if the corresponding functions are equal up to a permutation of V. Because any permutation of V is allowed, the group acting on V is S_n, $n = |V|$. This group induces an action on $[V]^2$ called $[S_n]^2$. It is our goal to calculate $Z([S_n]^2)$.

Assume that $g \in S_n$ has cycle structure $c = (c(1), \ldots, c(n))$. We determine the cycle structure of g as a permutation of $[V]^2$ by assuming $\{x, y\} \in [V]^2$ and examining four cases. To avoid confusion, we call the cycles in $[V]^2$ *pair-cycles*.

1. Suppose x and y lie in cycles of different lengths a and b. There are $c(a)$ choices for which cycle contains x and $c(b)$ choices for which cycle contains y. Once these choices are made the ab ordered pairs in the Cartesian product of the two cycles are partitioned into pair-cycles of length lcm $\{a, b\}$. The number of such cycles is $ab/\text{lcm}\{a, b\} = \gcd\{a, b\}$. The contribution to $Z([S_n]^2)$ is

$$\prod_{1 \le a < b \le n} x_{\text{lcm}\{a,b\}}^{c(a)c(b)\gcd\{a,b\}} . \tag{8.14}$$

2. Suppose x and y lie in different cycles of the same length a. There are $\binom{c(a)}{2}$ choices for the cycles. The pair-cycle has length a, and the number of pair-cycles is a. The contribution to $Z([S_n]^2)$ is

$$\prod_{1 \le a \le n} x_a^{a\binom{c(a)}{2}} . \tag{8.15}$$

3. Suppose x and y lie in the same cycle of odd length a. There are $c(a)$ choices for the cycle containing x and y and there are $(a-1)/2$ choices for the gap between x and y. The pair-cycles all have length a. The contribution to $Z([S_n])^2$ is

$$\prod_{a \text{ odd}} x_a^{\frac{c(a)(a-1)}{2}} . \tag{8.16}$$

4. Suppose x and y lie in the same cycle of even length a. This case is the same as Case 3 except for one important difference. There are still $c(a)$ choices for the cycle containing x and y. But now, although the typical pair-cycle has length a (and there are $(a-2)/2$ choices for the gap between x and y), there is also the possibility that x and y are $a/2$ units apart in the cycle, so that the pair-cycle has length $a/2$. The contribution to $Z([S_n])^2$ is

$$\prod_{a \text{ even}} x_a^{\frac{c(a)(a-2)}{2}} x_{a/2}^{c(a)} . \tag{8.17}$$

Putting (8.14), (8.15), (8.16), and (8.17) together we obtain $Z([S_n]^2) =$

$$\frac{1}{n!} \sum_c h(c) \prod_{1 \le a < b \le n} x_{\text{lcm}\{a,b\}}^{c(a)c(b)\gcd\{a,b\}} \prod_{1 \le a \le n} x_a^{a\binom{c(a)}{2}} \prod_{a \text{ odd}} x_a^{\frac{c(a)(a-1)}{2}}$$

$$\prod_{a \text{ even}} x_a^{\frac{c(a)(a-2)}{2}} x_{a/2}^{c(a)} . \tag{8.18}$$

The cycle indexes for $2 \le n \le 4$ are easily calculated.

$$Z([S_2]^2) = \tfrac{1}{2}(x_1^2 + x_2) \tag{8.19}$$

$$Z([S_3]^2) = \tfrac{1}{6}(x_1^3 + 3x_1x_2 + 2x_3) \tag{8.20}$$

$$Z([S_4]^2) = \tfrac{1}{24}(x_1^6 + 9x_1^2x_2^2 + 8x_3^2 + 6x_4x_2). \tag{8.21}$$

We apply Pólya's theorem to the cycle index $Z([S_4]^2)$ to enumerate nonisomorphic graphs of order 4 by number of edges:

$$Z([S_4]^2)[x_i \leftarrow y_1^i + y_2^i] = y_1^6 + y_1^5y_2 + 2y_1^4y_2^2 + 3y_1^3y_2^3 + 2y_1^2y_2^4 \\ + y_1y_2^5 + y_2^6. \tag{8.22}$$

According to Pólya's theorem, the coefficient of $y_1^a y_2^{6-a}$ in this enumerating polynomial is the number of nonisomorphic graphs of order 4 with a edges and $6 - a$ non-edges. The symmetry in the polynomial comes from the fact that G and H are isomorphic if and only if their complements G^c and H^c are isomorphic.

We verify (8.22) by presenting in Figure 8.1 all nonisomorphic graphs of order 4, arranged by number of edges.

Of course, the total number of nonisomorphic graphs of order 4 is just $Z([S_4]^2)[x_i \leftarrow 2] = Z([S_4]^2)[y_1 \leftarrow 1, y_2 \leftarrow 1] = 1 + 1 + 2 + 3 + 2 + 1 + 1 = 11$.

Let $g(n)$ be the number of nonisomorphic graphs on n vertices. Table 8.1 gives the value of $g(n)$ for small n. It can be shown that

$$g(n) \sim \frac{2^{\binom{n}{2}}}{n!}.$$

Table 8.1 The number of nonisomorphic graphs of small orders.

Order	Number of graphs
1	1
2	2
3	4
4	11
5	35
6	156
7	1,044
8	12,346
9	274,668
10	12,005,168

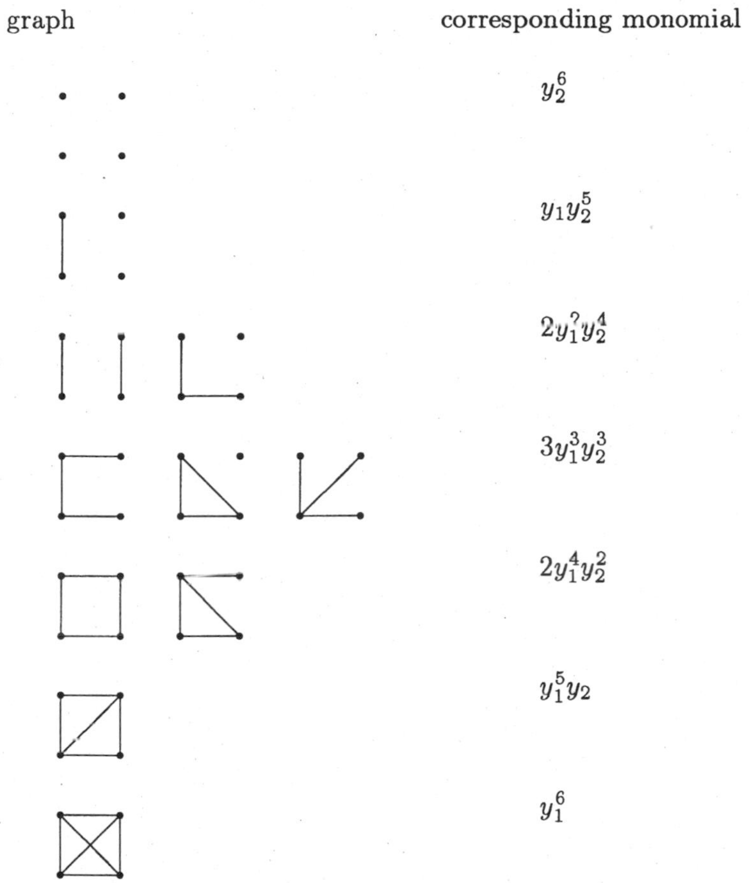

graph	corresponding monomial

$$y_2^6$$

$$y_1 y_2^5$$

$$2y_1^2 y_2^4$$

$$3y_1^3 y_2^3$$

$$2y_1^4 y_2^2$$

$$y_1^5 y_2$$

$$y_1^6$$

Figure 8.1. Graphs of order 4 and their enumerating monomials.

Because $2^{\binom{n}{2}}$ is the number of labeled graphs, this means that almost all graphs have no nontrivial automorphisms. It was an exercise in Chapter 4 to show that almost all labeled graphs are connected. In fact, $g(n) \sim c(n)$, where $c(n)$ is the number of nonisomorphic connected graphs of order n. See Harary and Palmer (1973).

Open Problem 16. *Find a formula for the number of nonisomorphic graphs of order n that contain a triangle.*

Open Problem 17. *Find a formula for the number of nonisomorphic planar graphs of order n.*

8.5.   DE BRUIJN'S FORMULA

Suppose X and Y are finite nonempty sets acted upon by finite groups G and H, respectively. We picture this situation with the following diagram:

$$f : \overset{G}{\underset{\downarrow}{X}} \longrightarrow \overset{H}{\underset{\downarrow}{Y}} .$$

We want to define what it means for two functions to be the same under these group actions, and, to this end, we define a new action H^G on the set of functions $Y^X = \{f : X \longrightarrow Y\}$ as follows: if $g \in G, h \in H$, and $f \in Y^X$, then $(h^g f)(x) = hf(g^{-1}x)$. (The reader should check that the two axioms for a group action are satisfied.) We say that two functions are *equivalent* if they are in the same orbit of the action and *inequivalent* otherwise. It should be noted that this definition is a generalization of the labeled/unlabeled set paradigm of Chapter 5. If G is the symmetric group S_x, then X is unlabeled, and if G is the identity group E_x, then X is labeled. Likewise, if H is S_y, then Y is unlabeled, and if H is E_y, then Y is labeled. For any groups G and H, Burnside's lemma gives the number of inequivalent functions:

$$N = \frac{1}{|G|}\frac{1}{|H|}\sum_{g \in G}\sum_{h \in H}\Omega(g,h), \qquad (8.23)$$

where $\Omega(g,h)$ is the number of functions fixed by h^g. If f is fixed by h^g, then $f(g^{-1}x) = hf(x)$ for all $x \in X$. Thus $f(x) = y$ implies $f(g^{-1}x) = hf(x) = hy$, which in turn implies $f(g^{-2}x) = h^2 y$. In general, $f(g^{-i}x) = h^i y$. It follows that if x is in a cycle of length i in g and y is in a cycle of length j in h, then j divides i. Furthermore, we have shown that the correspondence between the two cycles is completely determined by the equation $f(x) = y$. There are j choices for the image of x. Suppose g has cycle structure $(c(1), \ldots, c(m))$ and h has cycle structure $(d(1), \ldots, d(n))$. Then, given i and a particular cycle of length i in g, the number of fixed functions is

$$m_i(h) = \sum_{j|i} jd(j), \qquad (8.24)$$

and the total number of fixed functions is

$$\Omega(g,h) = \prod_i m_i(h)^{c(i)}. \tag{8.25}$$

Equations (8.23) through (8.25) combine to yield the following formula due to N. G. de Bruijn.

Theorem 8.4. *(De Bruijn's formula). If finite groups G and H act on finite nonempty sets X and Y, respectively, then the number of inequivalent functions is given by*

$$N = \frac{1}{|H|} \sum_{h \in H} Z(G)[x_i \leftarrow m_i(h)]. \tag{8.26}$$

We apply De Bruijn's formula to the problem of counting self-symmetric graphs of order n, that is, graphs G for which G is isomorphic to its complement G^c. In the previous section we determined the cycle index $Z([S_n]^2)$ and the number of nonisomorphic graphs of order n, $g_n = Z([S_n])[x_i \leftarrow 2]$. If $[S_n]^2$ acts on the set $X = \{1, \ldots, n\}$ and S_2 acts on $Y = \{0,1\}$, then functions correspond to nonisomorphic graphs where G is regarded the same as G^c. According to (8.26), the number of such functions is

$$N(n) = \frac{1}{2} \sum_{h \in S_2} Z([S_n]^2)[x_i \leftarrow m_i(h)]. \tag{8.27}$$

Because $2N(n)$ counts nonisomorphic graphs with each self-complementary graph counted twice, we let sc_n be the number of self-complementary graphs of order n and arrive at the formula

$$sc_n = 2N(n) - g_n. \tag{8.28}$$

For example, $sc_n = 2N(4) - g_4 = 2 \cdot 6 - 11 = 1$, and the unique self-complementary graph of order 4 is the path P_4.

Notes

In 1937 G. Pólya published the formula for enumerating graphs in connection with a problem concerning the number of chemical isomers. The language of his counting theory was quite descriptive. A function from X to Y is called a *configuration*. The elements of X are *places* and the elements of Y are *figures*. The *store inventory* is $\sum_{j=1}^n y_j^i$ and the *pattern inventory* is $Z(G)[x_1 \longleftarrow \sum_{j=1}^{n} y_j^i]$. Thus the basic problem is to find the number of G-inequivalent patterns. An alternative approach to the Pólya theory of counting was undertaken independently by J. H. Redfield in 1927.

Exercises

8.1 How many conjugacy classes have S_5? Let $x = (1\ 2\ 3)(4)(5)$. Find $|\mathrm{ccl}(x)|$ and $C(x)$.

8.2 Let $\mathcal{Z}_4$ rotate a cube around an axis passing through the centers of opposite faces. Verify Burnside's lemma for
X = the set of vertices of the cube,
Y = the set of faces of the cube, and
Z = the set of edges of the cube.

8.3 What is the average number of 1-cycles in the group A_n? What about in D_n?

8.4 Calculate $Z(\mathcal{Z}_3)$, $Z(D_3)$, $Z(A_3)$, and $Z(S_3)$.

8.5 How many 5-bead necklaces may be made using two types of beads? How many with three types of beads?

8.6 Let $Z(S_0) = 1$ and prove the identity $\sum_{n=0}^{\infty} Z(S_n) = \exp \sum_{j=1}^{\infty} (x_j/j)$. Recalling Exercise 5.4, let a_n be the number of functions $f : X \longrightarrow X$, $|X| = n$, with the property that $f(f(x)) = x$ for all $x \in X$. Show that the exponential generating function for the sequence $\{a_n\}$ is $\sum_{n=0}^{\infty} (a_n x^n/n!) = \exp[(x + x^2)/2]$. Note that a_n is also equal to the number of $n \times n$ symmetric permutation matrices.

8.7 How many ways may the 8 vertices of a cube be colored with 3 colors (up to symmetry of the cube)? How many ways may the 6 faces be colored? How many ways may the 12 edges be colored?

8.8 How many stacks of 10 poker chips contain 4 black chips and 6 white chips (up to inversion of the stack)?

8.9 How many necklaces of 10 beads may be made from two types of beads? How many with 5 of each type?

8.10 How many ways may 8 identical markers be placed on an 8×8 square grid (up to a rotation of the grid)?

8.11 How many ways may the 12 faces of a regular dodecahedron be colored with two colors? How many ways with 6 faces of each color?

8.12 Write the generating function for nonisomorphic multigraphs in terms of $Z(S_n)$. A multigraph is a graph in which pairs of vertices can be connected by more than one edge.

8.13 How many nonisomorphic multigraphs of order 4 have at most 5 edges?

8.14 How many self-complementary graphs are there of orders 5 and 6?

PART

III

Construction

Sixteen 7-dimensional unit hyperspheres can be placed at the vertices of the 7-dimensional unit hypercube so that each vertex of the hypercube lies on or inside exactly one hypersphere.

This configuration of spheres is called a *perfect packing*. How do we arrive at such a packing? What makes it perfect? What are its combinatorial properties? In Part III we draw on our experience with existential and enumerative combinatorics to address these questions as they pertain to packings and related constructions, including codes, geometries, t-designs, and Latin squares. We pay close attention to the interrelationships of these constructions, often finding equivalences between seemingly different structures. As a capstone, we construct the $(23, 2^{12}, 7)$ Golay code G_{23}, the $S(5, 8, 24)$ Steiner system, and Leech's 24-dimensional lattice L.

9

Codes

The science of coding theory is much too vast to be covered in one book, as it contains the studies of information transmission, error-correcting codes, and cryptology. In Part III we cover only the most straightforward aspects of perfect error-correcting codes and their related combinatorial designs. The reader can obtain background on information transmission over noiseless and noisy channels from McEliece (1977). For a general introduction to error-correcting codes, a good source is Pless (1989). For cryptology, Van Tilborg (1988) is recommended. We presume none of this material. Instead, we introduce codes as subsets of $GF(2)^n$ and take for granted that the fundamental problem concerning binary error-correcting codes is to find codes with high rate and large minimum Hamming distance.

9.1. THE GEOMETRY OF $GF(2)^n$

Let $F = GF(2)$, the field of two elements 0 and 1, and let F^n be the vector space of n-tuples of elements of F. A vector $v \in F^n$ is represented by a string of length n over F, e.g., $v = 010101 \in F^6$. Clearly, F^n has cardinality 2^n.

We sometimes picture F^n as the set of vertices of the n-dimensional unit hypercube C^n. For example, Figure 9.1 depicts F^3 as the set of vertices of C^3. These vertices are coordinatized with the eight vector representatives 000, 001, 010, 011, 100, 101, 110, and 111. Note that two vertices are edge-adjacent if and only if their vector representatives differ in exactly one coordinate.

This hypercube representation allows us to define a geometry (via a metric) on F^n. We denote by $d(v, w)$ the smallest number of edges of C^n in a path from v to w. For example, we can see in Figure 9.1 that $d(011, 101) = 2$. Equivalently, $d(v, w)$ is the number of components in which v and w differ. The following theorem shows that the function d is a metric, which we call the *Hamming metric* or *Hamming distance*.

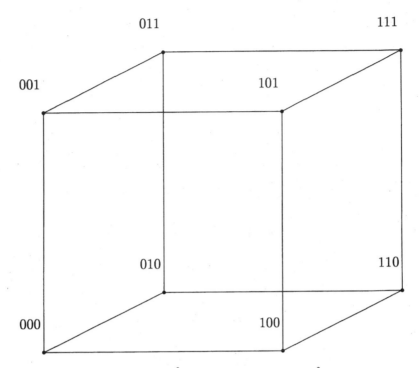

Figure 9.1. F^3 as the set of vertices of C^3.

Theorem 9.1. *The function* d *is a metric on* F^n; *that is, for all* $v, w, x \in F^n$ *the following properties hold:*

(1) $d(v, w) \geq 0$, *with equality only when* $v = w$ *(positivity).*

(2) $d(w, v) = d(v, w)$ *(symmetry).*

(3) $d(v, w) + d(w, x) \geq d(v, x)$ *(triangle inequality).*

Proof. Properties (1) and (2) are immediate from the definition of d. We prove the triangle inequality by verifying that it is preserved componentwise. Let v_i, w_i, x_i be the ith components of the vectors v, w, x. If $v_i = w_i = x_i$, then the contribution to both sides of the inequality is 0. If not, then the contribution to the left side is at least 1 while the contribution to the right side is at most 1. Therefore, the inequality is preserved componentwise. $\square$

To illustrate the dependence of geometry on the underlying metric, we describe two metrics on $\mathcal{R}^2$, one familiar and the other perhaps not so familiar. Given two points $x = (x_1, x_2)$ and $y = (y_1, y_2)$ in $\mathcal{R}^2$, the usual *Euclidean distance* between x and y is given by $d_1(x, y) = ((x_1 - y_1)^2 + (x_2 - y_2)^2)^{1/2}$. We leave it to the reader to check that the three conditions for a metric are met. The *circle* $O(c, r)$ with center $c = (c_1, c_2)$ and radius r is defined as

$$O(c, r) = \{x \in \mathcal{R}^2 : d_1(x, c) = r\}$$
$$= \{(x_1, x_2) : (x_1 - c_1)^2 + (x_2 - c_2)^2 = r^2\}.$$

For instance, the graph of $O(0, 4)$ is a circle with radius 4 centered at the origin, whose equation is

$$x_1^2 + x_2^2 = 16.$$

Alternatively, we define a different distance, called the *taxicab distance*, between x and y, by $d_2(x, y) = |x_1 - y_1| + |x_2 - y_2|$. If x and y are located at street intersections of a city with streets running north–south and east–west, then $d_2(x, y)$ is the distance a taxicab traverses driving from x to y. If we define circles as we did for the Euclidean metric, then $O(0, 4)$ is the set of points (x_1, x_2) for which

$$|x_1| + |x_2| = 4.$$

Surprisingly, the graph of this "circle" is a square. Figure 9.2 illustrates the circle $O(0, 4)$ with respect to the two metrics d_1 and d_2.

In F^n endowed with the Hamming metric, we define the *sphere* $S(c, r)$ with *center* c and *radius* r as the set $S(c, r) = \{v \in F^n : d(c, v) \leq r\}$. (Our definition of sphere includes the interior.) The *volume* of a sphere is its cardinality; that is,

$$volume(S(c, r)) = |S(c, r)|$$
$$= \sum_{k=0}^{r} |\{v \in F^n : d(c, v) = k\}|$$
$$= \sum_{k=0}^{r} \binom{n}{k}.$$

In the last summation, $\binom{n}{k}$ counts selections of the k components in which c and v disagree.

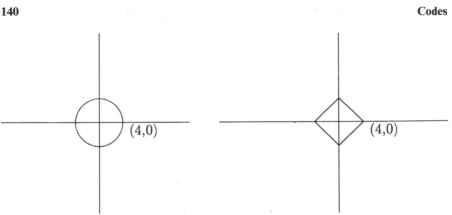

Figure 9.2. $O(0,4)$ with respect to metrics d_1 (left) and d_2 (right).

It is difficult to picture spheres when n is large, and they don't look very spherical. We will mainly be interested in how densely they can be packed, because, as we shall see, dense packings signify good codes.

9.2.   BINARY CODES

A *code* A is a subset of F^n with $|A| \geq 2$. The elements of A are called *codewords*. In real-life applications, information can be sent reliably over a noisy channel by encoding redundancy in the message. A codeword $v \in A$ is transmitted and a possibly distorted vector v' is received. As it might happen that v' equals a codeword in A different from v, it is not always possible to tell whether any errors have been committed in the transmission. However, if the Hamming distances between pairs of codewords are fairly large (which is possible with redundancy), it is unlikely that v' will equal another codeword. If the Hamming distances are large enough, it may even be possible to determine where the errors have occurred and correct them.

The *distance* $d(A)$ of a code A is the minimum Hamming distance between distinct codewords in A. For example, the code $A = \{011, 101, 110\} \subseteq F^3$ has distance $d(A) = 2$, because any two of the vectors in A differ by two bits. Always, $1 \leq d(A) \leq n$.

If $d(A) \geq e + 1$, then we say that A *detects* e errors, and if $d(A) \geq 2e + 1$, then A *corrects* e errors. To justify these definitions, suppose a codeword v is sent and a string v' is received, with $1 \leq d(v, v') \leq e$. If $d(A) \geq e + 1$, then v' cannot equal some codeword x or else $d(v, v') \geq e + 1$, a contradiction. Therefore, we can detect that errors have occurred. If $d(A) \geq 2e + 1$, then v' cannot have resulted from the transmission of an erroneous codeword x (and at most e

errors), or else $2e + 1 \leq d(v, x) \leq d(v, v') + d(v', x) \leq e + e = 2e$, a contradiction. Therefore, we can identify the particular vector v that was sent and correct the errors.

For example, suppose $A = \{000, 111\}$, a code with distance 3 capable of correcting 1 error. If a codeword is transmitted and we receive 001, then, under the assumption that at most one error has been committed, the intended word must be 000. Let q be the probability that a bit is mistakenly altered from 0 to 1 or from 1 to 0, and suppose that bits are altered independently of one another. The decoding scheme fails if 2 or 3 errors occur, and this happens with probability $3q^2(1 - q) + q^3$, which is asymptotic to $3q^2$ for q small. This probability of failure is much smaller than the probability q of failure when no code is used. However, the increase in reliability is paid for by a decrease in transmission rate. The *rate* $r(A)$ of a code A, defined as $r(A) = (\log_2 |A|)/n$, measures the amount of information per code bit conveyed over the communication channel. Always, $1/n \leq r(A) \leq 1$. In the above example, $r(A) = (\log_2 2)/3 = \frac{1}{3}$, which means that when information is sent in triplicate the rate decreases by a factor of 3 (which is reasonable). For more information about the numerical aspects of coding, the reader should consult McEliece (1977).

We refer to a code $A \subseteq F^n$ with distance $d(A) = d$ as an $(n, |A|, d)$ code. The numbers n, $|A|$, and d are called *parameters*. The number n is sometimes called the *dimension* of the code, and $|A|$ is called the *size* of the code. Let there be no confusion between the distance d (an integer) and the Hamming metric d (a function). For n fixed, there is an inverse relationship between $|A|$ and d (and therefore between the rate and the error-correcting capability of the code). The *fundamental problem of coding theory* is to produce codes which maximize $|A|$ and d simultaneously. The next theorem gives sharp focus to this problem.

Theorem 9.2.　　*(Hamming upper bound). If A corrects e errors, then*

$$|A| \leq \frac{2^n}{\sum_{k=0}^{e} \binom{n}{k}}. \tag{9.1}$$

Proof.　　Because $d \geq 2e + 1$, the spheres $S(v, e)$, for $v \in A$, do not intersect. Therefore, the total volume of spheres is at most the cardinality of F^n; that is,

$$2^n \geq \sum_{v \in A} |S(v, e)|$$

$$= |A| \sum_{k=0}^{e} \binom{n}{k},$$

from which (9.1) follows instantly. $\square$

The Hamming upper bound for $|A|$ is also called the *sphere packing bound*.

9.3. PERFECT CODES

If the Hamming upper bound is achieved for a code A, we say that A is *perfect*. Perfect codes correspond to sphere packings of F^n with no wasted space (vectors not in a sphere). If A is perfect, then $\sum_{k=0}^{e} \binom{n}{k}$ must divide 2^n, and hence be a power of 2. We will soon see that this rarely happens.

If $|A| = 2$, then the maximum value of $d(A)$ is n, and this value is achieved, for instance, when A consists of the all 0 vector and the all 1 vector. If $|A| > 2$, then some two vectors must agree on any given component, so $d(A) < n$. Therefore, $n > d \geq 2e + 1$, which implies

$$e < \frac{n-1}{2}. \tag{9.2}$$

If $e = 1$, then $\sum_{k=0}^{e} \binom{n}{k} = \binom{n}{0} + \binom{n}{1} = n + 1$, which is a power of 2 when $n = 2^r - 1$ for some r. In this case, $|A| = \frac{2^n}{2^r} = 2^{n-r} = 2^m$, where $m = n - r = 2^r - 1 - r$. Hence, the parameters of such a code are $(n, |A|, d) = (2^r - 1, 2^m, 3)$. To avoid trivialities we shall assume $r \geq 2$.

The next theorem says that there are only two feasible sets of parameters for perfect codes when $e > 1$. See Pless (1989).

Theorem 9.3. *The only values of n and e for which*

(1) $1 < e < \frac{n-1}{2}$ and

(2) $\sum_{k=0}^{e} \binom{n}{k}$ is a power of 2

are $(n, e) = (23, 3)$ and $(90, 2)$.

If there are codes with these values of n and e, they would have parameters $(n, |A|, d) = (23, 2^{12}, 7)$ and $(90, 2^{78}, 5)$. We outline a proof in the exercises that there is no code with parameters $(90, 2^{78}, 5)$. However, a code with parameters $(23, 2^{12}, 7)$, called the Golay code G_{23}, is constructed in Section 11.1.

We have taken the base field of our codes to be $GF(2)$, but if we allow other base fields it turns out that there is just one more perfect code, a ternary code G_{11} with parameters $(11, 3^6, 5)$, discovered by M. Golay. If we consider any alphabet as the base set (not necessarily a field), then a perfect code is one of the two Golay codes G_{23} and G_{11} or else has the parameters of a Hamming code.

In the next section, we describe a family of perfect 1-error correcting codes called Hamming codes.

9.4. HAMMING CODES

As in the previous section, we let $r \geq 2$ and define $n = 2^r - 1$ and $m = 2^r - 1 - r$. First we note that if $r = 2$, then $n = 3$ and $m = 1$, and the vertices 000 and 111 of the cube C^3 of Figure 9.1 constitute such a code. We now describe a code $A \subseteq F^n$ with $|A| = 2^m$ which corrects one error, exhibiting a construction in the case $r = 3, n = 7, m = 4$. (The constructions for higher values of r, n, m are carried out similarly.) Let

$$H = \begin{bmatrix} 1 & 0 & 1 & 0 & 1 & 0 & 1 \\ 0 & 1 & 1 & 0 & 0 & 1 & 1 \\ 0 & 0 & 0 & 1 & 1 & 1 & 1 \end{bmatrix}$$

be the $r \times n = 3 \times 7$ matrix whose columns are the numbers $1, \dots, n$ written in binary. The matrix H represents a linear transformation from F^7 to F^3:

$$H : F^7 \longrightarrow F^3$$

$$v \mapsto Hv.$$

We define the code A to be the kernel of H; that is,

$$A = \{v \in F^7 : Hv = 0\}.$$

(The 0 here is the 3×1 all zero vector.) We call H a *parity check matrix* for A. Any code A for which $x \in A$ and $y \in A$ imply $x + y \in A$ is called a *linear* code. Clearly, a code described as the kernel of a parity check matrix is a linear code.

By inspection, we find two of the vectors belonging to A:

$$\begin{bmatrix} 0 \\ 0 \\ 0 \\ 0 \\ 0 \\ 0 \\ 0 \end{bmatrix} \quad \text{and} \quad \begin{bmatrix} 1 \\ 1 \\ 1 \\ 0 \\ 0 \\ 0 \\ 0 \end{bmatrix}.$$

As the Hamming distance between these two vectors is 3, we see that $d(A) \leq 3$. We need to prove that $d(A) \geq 3$ and $|A| = 2^m = 16$.

Assume $v = [x\,y\,a\,z\,b\,c\,d]^t \in A$. The reason for the nonalphabetical listing of the components of v will become clear in a moment. Because v is in the kernel of H, $Hv = 0$. Thus

$$\begin{bmatrix} 1 & 0 & 1 & 0 & 1 & 0 & 1 \\ 0 & 1 & 1 & 0 & 0 & 1 & 1 \\ 0 & 0 & 0 & 1 & 1 & 1 & 1 \end{bmatrix} \begin{bmatrix} x \\ y \\ a \\ z \\ b \\ c \\ d \end{bmatrix} = \begin{bmatrix} 0 \\ 0 \\ 0 \end{bmatrix},$$

which yields three equations

$$x + a + b + d = 0$$
$$y + a + c + d = 0$$
$$z + b + c + d = 0.$$

Because we are working in F, $-x = x$ for all x, and the equations become

$$x = a + b + d$$
$$y = a + c + d$$
$$z = b + c + d.$$

The variables a, b, c, d may independently take either value, 0 or 1, in F. For this reason they are called *free variables*. There are 2^4 choices for the values of the four free variables. The variables x, y, z are determined by these choices and are therefore called *determined variables*.

We have shown that $|A| = 16$. It remains to prove $d(A) \geq 3$, which we do by showing that A corrects one error. Suppose a codeword $c \in A$ is sent and one error occurs. Assuming the error occurs in the ith component, we represent the error by a vector consisting of a single 1 in the ith position and zeros in all other positions:

$$e = \begin{bmatrix} 0 \\ \vdots \\ 0 \\ 1 \\ 0 \\ \vdots \\ 0 \end{bmatrix} \longleftarrow i\text{th position}$$

The received vector is $c + e$ (which differs from c in just the ith component), and it is the decoder's job to determine the position in which the error has occurred. This is done by exploiting special properties of the matrix H. We multiply H by $c + e$:

$$H(c + e) \doteq Hc + He$$
$$= 0 + He \quad \text{(by definition of } A\text{)}$$
$$= He.$$

Because e has only one nonzero row, the product He consists of the column of H corresponding to this row position. In other words, He equals the ith column of H, which happens, because of the way H is constructed, to be the number i in binary. Thus when we compute $H(c + e)$ the position of the error is revealed (in binary). We have demonstrated that A corrects one error, so $d(A) \geq 2 \cdot 1 + 1 = 3$.

We have already remarked that A is a perfect code. As a sphere packing, A may be pictured as the centers of the spheres referred to in the introduction to Part III. As we have already noted, every element of F^3 lies in exactly one sphere. This Hamming code (of dimension 7) has rate $r(A) = (\log_2 16)/7 = \frac{4}{7}$. In general, the Hamming code of dimension n has rate

$$r(A) = \frac{\log_2 2^m}{n} = \frac{m}{n} = \frac{2^r - 1 - r}{2^r - 1},$$

which tends to 1 as r tends to infinity. Thus the Hamming codes are a family of 1-error correcting codes with arbitrarily good rate.

9.5.   THE FANO CONFIGURATION

The equations of the previous section allow us to list the sixteen codewords of the $(7, 16, 3)$ Hamming code. They are

$$\begin{bmatrix}0\\0\\0\\0\\0\\0\\0\end{bmatrix}\begin{bmatrix}1\\1\\0\\1\\0\\0\\1\end{bmatrix}\begin{bmatrix}0\\1\\0\\1\\0\\1\\0\end{bmatrix}\begin{bmatrix}1\\0\\0\\0\\0\\1\\1\end{bmatrix}\begin{bmatrix}1\\0\\0\\1\\1\\0\\0\end{bmatrix}\begin{bmatrix}0\\1\\0\\0\\1\\0\\1\end{bmatrix}\begin{bmatrix}1\\1\\0\\0\\1\\1\\0\end{bmatrix}\begin{bmatrix}0\\0\\0\\1\\1\\1\\1\end{bmatrix}$$

$$\begin{bmatrix}1\\1\\1\\0\\0\\0\\0\end{bmatrix}\begin{bmatrix}0\\0\\1\\1\\0\\0\\1\end{bmatrix}\begin{bmatrix}1\\0\\1\\1\\0\\1\\0\end{bmatrix}\begin{bmatrix}0\\1\\1\\0\\0\\1\\1\end{bmatrix}\begin{bmatrix}0\\1\\1\\1\\1\\0\\0\end{bmatrix}\begin{bmatrix}1\\0\\1\\0\\1\\0\\1\end{bmatrix}\begin{bmatrix}0\\0\\1\\0\\1\\1\\0\end{bmatrix}\begin{bmatrix}1\\1\\1\\1\\1\\1\\1\end{bmatrix}.$$

The *weight* $w(v)$ of a vector v is the number of 1's in v. A simple componentwise proof demonstrates that $d(a, b) = w(a - b)$ for any $a, b \in F^n$. It follows that the distance d of a linear code equals the minimum weight of a nonzero codeword. Table 9.1 tallies the words of the $(7, 16, 3)$ Hamming code according to weight.

The symmetry of the weight distribution is due to the fact that A is a linear code containing the all 1 codeword. Thus, if $v \in A$, then also $v^c \in A$ (where v^c is the binary complement of v); and $w(v^c) = 7 - w(v)$.

Table 9.1 Weight distribution of the Hamming code.

weight	0	3	4	7
number of words	1	7	7	1

The seven Hamming codewords of weight 3 give rise to an interesting finite geometry called the *Fano configuration* (*FC*), first considered by the geometer G. Fano (1871–1952). The Fano configuration has seven points, 1, 2, 3, 4, 5, 6, 7, corresponding to the seven components of the code vectors of *A*. Each codeword of weight 3 contains, by definition, three 1's. The three points corresponding to the 1's are joined by a line, called an *edge* of *FC*. The edges are: 246, 167, 145, 257, 123, 347, and 356. Figure 9.3 shows one way of drawing *FC*. However, we must emphasize that the lines are just triples of points; they have no Euclidean meaning. Therefore, the points may be located anywhere in the plane, and the lines may be drawn straight or curved and may cross arbitrarily.

We note that *FC* has the following properties:

1. There are seven points.
2. There are seven lines.
3. Every line is incident with three points.
4. Every point is incident with three lines.
5. Each pair of lines determines a unique point.
6. Every two points determine a unique line.

The above properties occur in pairs called *duals*. If the words 'point' and 'line' are interchanged, each property is transformed into its dual property. We describe other combinatorial objects with dual structure in Chapter 10.

As the prototype of many families of combinatorial structures, *FC* abounds with fascinating properties. For instance, with its edges properly oriented, *FC* represents the multiplication table for the Cayley algebra. We investigate some generalizations of *FC* in Chapter 10. Here we content ourselves with determining Aut *FC*, the *automorphism*

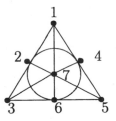

Figure 9.3. The Fano configuration *FC*.

group of permutations of the vertices of *FC* which preserve collinearity of points.

We calculate |Aut *FC*|, the order of the automorphism group of *FC*. It is evident from Figure 9.3 that all seven vertices are equivalent in terms of collinearity. Therefore, vertex 1 may be sent by an automorphism to any of the seven vertices. Suppose 1 is mapped to $1'$. Vertex 2 may be mapped to any of the remaining six vertices. Suppose 2 is mapped to $2'$. In order to preserve collinearity, vertex 3 must be mapped to the unique point collinear with $1'$ and $2'$. Call this point $3'$. Vertex 4 is not on line 123, so its image $4'$ can be any of the remaining four vertices. Finally, the images of the other points are all determined by collinearity: 5 is collinear with 1 and 4; 6 is collinear with 2 and 4; and 7 is collinear with 3 and 4. Hence, there are $7 \cdot 6 \cdot 4 = 168$ automorphisms.

Now we know that Aut *FC* is a group of order 168, but what group? Is it abelian? cyclic? simple? It turns out that Aut *FC* is isomorphic to a group of matrices over a finite field; but first a little background. The *general linear group* $GL(n, q)$ is the set of invertible $n \times n$ matrices with coefficients in the Galois field $GF(q)$ of order $q = p^k$, where p is prime, under matrix multiplication. The order of $GL(n, q)$ is readily determined. There are $q^n - 1$ choices for the first row of an invertible $n \times n$ matrix (the all zero row is excluded). Having chosen the first row, the second row may be any of the q^n possible n-tuples except the q scalar multiples of the first row. Thus, there are $q^n - q$ choices for the second row. Similarly, the third row may be any of the q^n n-tuples except linear combinations of the first two rows. There are $q^n - q^2$ choices. Continuing in this manner, we arrive at the total number of invertible matrices:

$$|GL(n, q)| = (q^n - 1)(q^n - q)(q^n - q^2) \cdots (q^n - q^{n-1}). \quad (9.3)$$

For example, $GL(2, 2)$ has 6 elements and is in fact isomorphic to S_3.

The *special linear group* $SL(n, q)$ is a normal subgroup of $GL(n, q)$ consisting of those $n \times n$ invertible matrices with entries from $GF(q)$ and determinant 1. We claim that $SL(n, q)$ is a normal subgroup of $GL(n, q)$. For if $M \in SL(n, q)$ and $N \in GL(n, q)$, then $\det(NMN^{-1}) = \det N \det M \det N^{-1} = \det N \det N^{-1} = \det(NN^{-1}) = \det I = 1$. Now consider the homomorphism

$$f : GL(n, q) \longrightarrow GF(q) - \{0\}$$
$$M \mapsto \det M.$$

The kernel of f is $SL(n,q)$ and the homomorphism is clearly onto. Therefore, by the first homomorphism theorem for groups, $GL(n,q)/SL(n,q) \simeq GF(q) - \{0\}$, and hence, $|GL(n,q)|/|SL(n,q)| = q - 1$. Combining this equation with formula (9.3), we obtain a formula for $|SL(n,q)|$:

$$|SL(n,q)| = \frac{(q^n - 1)(q^n - q)(q^n - q^2) \cdots (q^n - q^{n-1})}{(q-1)}. \qquad (9.4)$$

We now mention projective versions of $GL(n,q)$ and $SL(n,q)$. The *projective general linear group* $PGL(n,q)$ is defined to be $GL(n,q)/Z(GL(n,q))$ and the *projective special linear group* $PSL(n,q)$ is defined to be $SL(n,q)/Z(SL(n,q))$, where $Z(H)$ denotes the center of H. We leave it to the reader to find formulas for the orders of these groups. We note that

$$GL(n,2) \simeq SL(n,2) \simeq PGL(n,2) \simeq PSL(n,2)$$

for all n. When $n = 3$ we obtain

$$|GL(3,2)| = \frac{(2^3 - 1)(2^3 - 2)(2^3 - 2^2)}{2 - 1} = 7 \cdot 6 \cdot 4 = 168.$$

We have shown that $GL(3,2)$ and Aut FC have the same order. Now we prove that the two groups are isomorphic.

We label the vertices of FC with the seven nonzero vectors in F^3, as in Figure 9.4. This labeling is derived from the labeling of Figure 9.3 by assigning to each vertex i the vector which represents the number i in binary. The vectors have been chosen so that v_1, v_2, v_3 are collinear if and only if $v_1 + v_2 + v_3 = 0$ (in F^3). The matrix group $GL(3,2)$ acts on the vectors of FC in the obvious way: $v \longmapsto Mv$. It is easy to check that this action preserves collinearity: $v_1 + v_2 + v_3 = 0$ if and only if

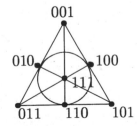

Figure 9.4. Vector representation of *FC*.

$M(v_1 + v_2 + v_3) = M \cdot 0$, which is true if and only if $Mv_1 + Mv_2 + Mv_3 = 0$. Therefore, $GL(3, 2)$ is isomorphic to FC.

It is well known (see Rotman (1973)) that $PSL(n, q)$ is a simple group (a noncyclic group with no nontrivial normal subgroups) for all $n \geq 2$ except $n = q = 2$. Because there is only one nonabelian group of order six, it follows that $PSL(2, 2) \simeq S_3$. The simple groups are crucial to the study of algebra. Some of them are difficult to describe and work with, although we have shown that $PSL(3, 2)$, at least, has a nice geometric model.

By direct calculation it can be shown that the columns of M are the images of the binary representations of 1, 2, and 4, respectively. Let

$$A = \begin{bmatrix} 1 & 0 & 0 \\ 0 & 0 & 1 \\ 0 & 1 & 0 \end{bmatrix} \quad \text{and} \quad B = \begin{bmatrix} 0 & 1 & 1 \\ 1 & 0 & 0 \\ 1 & 0 & 1 \end{bmatrix}.$$

We see from Figure 9.3 that A yields a flip of FC around the axis 176 while B yields the 7-cycle (1 6 4 5 3 7 2). By inspection we find that BA has order 3 (but isn't a rotation) and $B^4 A$ has order 4. Indeed, it can be shown (see Rotman (1973)) that $GL(3, 2)$ has the group presentation

$$\langle A, B : B^7 = A^2 = (BA)^3 = (B^4 A)^4 = 1 \rangle.$$

Notes

R. Hamming developed the Hamming codes in 1947 at Bell Telephone Laboratories in order to solve the problem of glitches in running computer programs. M. Golay did much of the same work independently at the Signal Corps Engineering Laboratories. For a history of their achievements the reader is referred to Thompson (1983). Today, error-correcting codes are used in the design of compact discs (CDs), satellite communications, ISBN numbers, and bar code scanners.

Error-correcting codes are also a natural feature of the science of information theory introduced by C. Shannon in 1941 (also at Bell Labs). The main ideas of information theory are that an information source contains a certain amount of uncertainty (entropy) and that entropy determines the accuracy and amount of information that can be sent over a communication channel. In a memoryless source of English characters, for example, the entropy is thought to be about 4.03 bits (each revealed character conveys about 4.03 bits of information). As a base-line comparison, the entropy of a memoryless source of 27 equally likely symbols (26 letters and a space) is approximately 4.76 bits. (By the way, in a popular word-making board game the distribution of letter tiles gives an entropy value of about 4.32 bits.) However, if knowledge of the English language is used to predict the likelihood of future characters (the source has a memory), then the entropy is a number believed to be between 0.6 and 1.3 bits. Knowledge of the information structure of a language is useful in the design of machines which translate the spoken word into written text (continuous speech recognition).

We do not deal in this book with cryptology, but a brief description of recent developments may help the reader interested in pursuing that subject. In 1970 IBM introduced the Data Encryption Standard (DES) for use in guaranteeing the privacy of communications. DES is the most widely used symmetric scheme (i.e., a scheme in which the same key is used for both encryption and decryption).

In 1977 at MIT a new type of code was introduced by R. Rivest, A. Shamir, and L. Adelman (RSA). Their method consists of generating large prime numbers p and q (greater than 10^{200}), letting $m = \phi(n) = (p-1)(q-1)$, and choosing a number x (at random) such that $\gcd(x, m) = 1$ and a number y (using the Euclidean algorithm) satisfying $xy \equiv 1 \bmod m$. The numbers x and $n = pq$ are announced publicly, while p, q, and m are kept private. The public encryption algorithm, which allows anyone to send a message to the user, is the function $f : \mathcal{Z}_n \longrightarrow \mathcal{Z}_n$, defined by $f(a) = a^x$. The receiver receives $f(a) = b$ and applies the decryption function defined by $g(x) = b^y$. By Fermat's theorem (Section 1.5), $g(f(a)) = a^{xy} \equiv a^{k(p-1)(q-1)+1} \equiv a \bmod p$ and $\bmod q$. It follows that $g(f(a)) \equiv a \bmod n$. Thus g are f are inverse functions. Moreover, this scheme has the remarkable feature that knowing how to encode does not tell one how to decode. The obvious way of cracking the code is to determine p and q (i.e., factor n), but this is computationally difficult.

The RSA code has another nice feature. It is possible to send, along with the message, a "signature" which guarantees the identity of the person sending it. Suppose A, who has encoding function f_A and decoding function g_A, wishes to communicate with B, who has functions f_B and g_B. While g_A and g_B are kept private, f_A and f_A are known publicly. Along with the message M, A sends a signature S which is encoded as $f_B(g_A^{-1}(S))$. B receives this and applies f_B^{-1} to obtain $g_A^{-1}(M)$. Then B employs the encoding function f_A to obtain M. Because only A knows f_A, B has verified that the message has come from A.

Progress toward breaking the so-called "unbreakable" RSA code was made in 1988 when a team led by A. Lenstra of the University of Chicago and M. Manasse of the Digital Equipments Corporation's Systems Research Center in Palo Alto factored a number n on the order of 10^{100}, thus uncovering the secret primes $p \approx 10^{40}$ and $q \approx 10^{59}$. This work required the coordinated use of about 400 computers spread over three continents (North America, Europe, and Australia). By 1990, the Lenstra–Manasse team was able to factor an integer on the order of 10^{154}, and more recently larger numbers have been factored. It is now believed that the problem of breaking the RSA code is not NP-complete. Another drawback of RSA is that it is computationally costly to encrypt and decrypt large files. However, the RSA technology may be used in conjunction with efficient symmetric block algorithms such as IDEA (International Data Encryption Algorithm, invented in 1990 by X. Lai and J. L. Massey) or DES. One application is to PEM (privacy enhanced mail), which allows e-mail with digital signatures to be sent securely over public channels. Basically, the idea is to convey the key for an algorithm such as DES by encrypting it via an asymmetric algorithm such as the RSA code. Thus the recipient gets an encrypted file and an encrypted key. The key can be decrypted using the privately held inverse to the RSA code. Once the symmetric key is recovered, it is used to decrypt the main file.

Today, cryptology is used by governments to ensure the security of their secret communications, by cable television companies to prevent unauthorized use of their transmissions, and by banks to protect their customers' transactions. As the RSA code becomes less secure for greater values of n, the search is back on for a truly unbreakable code. Indeed, in 1985 N. R. Wagner and M. R. Magyarik proposed a code whose security is based on the unsolvability of the word problem in combinatorial group theory; but, of course, this and all future encryption schemes are always vulnerable to ingenious new methods of decryption.

Exercises

9.1 Prove that the maximum number $A(n, e)$ of codewords in an e-error correcting code in F^n satisfies the *Gilbert lower bound*

$$A(n, e) \geq \frac{2^n}{\sum_{j=0}^{2e} \binom{n}{j}}.$$

9.2 Show that there exists no code A in $GF(2)^{10}$ with 19 words and $d(A) = 5$.

9.3 Find a code A in $GF(2)^8$ with $d(A) = 4$ and $r(A) = \frac{\log_2 7}{7}$.

9.4 Find a code A in $GF(2)^8$ with $d(A) = 4$ and $r(A) = \frac{1}{2}$.

9.5 Let A be the $(15, 2^{11}, 4)$ Hamming code. Suppose $v \in A$ is sent, at most one error occurs, and

$$w = 101000000000000$$

is received. Find v.

9.6 Use a computer to verify that the only ordered pairs (n, e) with $2 \le e < (n-1)/2 \le 90$ and $\sum_{j=0}^{e} \binom{n}{j}$ a power of 2 are $(23, 3)$ and $(90, 2)$.

9.7 Prove that there is no $(90, 2^{78}, 5)$ code.
[Hint: Assume that there is such a code.] Without loss of generality, we can assume that the code contains the 0 vector (why?). The code, being perfect, corresponds to a sphere packing of F^{90} with 2^{78} spheres of radius 2. Let X be the set of weight 3 vectors in F^{90} which have 1's in the first two components. Show that X has 88 elements. How are the elements of X partitioned by spheres around codewords of weight 5?

9.8 Find a design, other than the Fano configuration, with seven points and seven lines, each line containing three points, and each point on three lines.

9.9 (Sylvester's problem). Suppose A is a finite set of points in the plane with the property that every line determined by two points of A contains a third point of A. Prove that A is collinear.
[Note: This result shows that FC may not be drawn in the plane with straight lines.]

9.10 Show that the assertion in the previous exercise is false if A is infinite.

10

Designs

The Fano configuration FC of the previous chapter is the simplest nontrivial example of many types of combinatorial configurations, including t-designs, Steiner systems, block designs, and projective planes. In this chapter we explore these designs and investigate their interconnections, paving the way for the production in Chapter 11 of the Golay code G_{23}, the only perfect binary code capable of correcting more than one error.

10.1. t-DESIGNS

A t-(v, k, λ) *design* (or *t-design*) is an ordered pair (S, C) consisting of a v-set S and a collection C of k-subsets of S, with the property that every t-subset of S is contained in exactly λ elements of C. The elements of C are called *blocks*. A *nontrivial t*-design has $1 < t < k < v$ and not every k-subset of S is a block.

Example. The edges 246, 167, 145, 257, 123, 347, and 356 of the Fano configuration FC are the blocks of a 2-$(7, 3, 1)$ design with $S = \{1, \ldots, 7\}$.

Example. The complement FC' of FC has the same set of vertices as FC. Its edges are the complements of the edges of FC: 1357, 2345, 2367, 1346, 4567, 1256, 1247. It is easy to verify that the edges of FC' constitute the blocks of a 2-$(7, 4, 2)$ design.

Example. The following sets are the blocks of a 3-$(8, 4, 1)$ design.

1357	2345	2367	1346	4567	1256	1247
2468	1678	1458	2578	1238	3478	3568

These blocks are of two types: (1) the edges of FC', and (2) the edges of FC joined to a new element 8. The *derived* design obtained by removing

154

any point and all the sets not incident with it is equivalent to FC. Conversely, we call the 3-(8, 4, 1) design an *extension* of FC.

Examples. A graph with p vertices and q edges is a 0-$(p, 2, q)$ design. An r-regular graph is a 1-$(p, 2, r)$ design. An r-regular k-uniform hypergraph is a 1-(p, k, r) design.

We say that two *t*-designs are *equivalent* if they can be made equal by properly relabeling their underlying sets S_1 and S_2. Each of the designs above is unique up to this equivalence. One reason for the uniqueness is that many parameters of a design are determined by the following theorem.

Theorem 10.1. *(Parameter theorem). If (S, C) is a t-(v, k, λ) design, and $0 \le i \le t$, then there exists a constant λ_i such that every i-set of S lies in exactly λ_i elements of C. Therefore, (S, C) is also an i-(v, k, λ_i) design. Furthermore, λ_i satisfies $\lambda_i \binom{k-i}{t-i} = \lambda \binom{v-i}{t-i}$.*

Proof. Let X be a fixed subset of S with $|X| = i$, and consider the ordered pairs (T, K) with $|T| = t$, $K \in C$ and $X \subseteq T \subseteq K$. As in the proof of Burnside's lemma (Theorem 8.1), we count the ordered pairs in two ways (from the perspective of each coordinate) to obtain $\lambda_i \binom{k-i}{t-i} = \lambda \binom{v-i}{t-i}$, an equation independent of X. $\sqcup$

We solve the parameter equation for λ_i:

$$\lambda_i = \lambda \frac{\binom{v-i}{t-i}}{\binom{k-i}{t-i}}. \tag{10.1}$$

Putting in the value $i = 0$, we obtain the number b of blocks in C.

$$b = \lambda_0 = \lambda \frac{\binom{v}{t}}{\binom{k}{t}} \tag{10.2}$$

Putting in the value $i = 1$, we obtain the number r of times each element of S occurs in a block.

$$r = \lambda_1 = \lambda \frac{\binom{v-1}{t-1}}{\binom{k-1}{t-1}} \tag{10.3}$$

The reader should verify the formulas for b and r in the above examples.

An $S(t, k, v)$ *Steiner system* is a *t*-design with $\lambda = 1$. In Section 11.1 we construct the Golay code G_{23} via a related code which contains an

$S(5,8,24)$ Steiner system. No Steiner system is known with $t > 5$. The only known Steiner systems with $t = 5$ are $S(5,6,12)$, $S(5,8,24)$, $S(5,6,24)$, $S(5,6,48)$, $S(5,6,84)$, $S(5,7,28)$, and $S(5,6,72)$ designs, and the only known Steiner systems with $t = 4$ are derived from these designs. A *Steiner triple system* is a Steiner system with $k = 3$. For example, FC is a $S(2,3,7)$ Steiner triple system. An $S(2,3,9)$ Steiner triple system is given by the set of blocks $\{123, 456, 789, 147, 168, 159, 258, 369, 249, 357, 267, 348\}$. This Steiner system is equivalent to the set of nonideal points and lines of the projective plane of ordered 3 (Figure 10.1); see Section 10.3.

Open Problem 18 *Determine whether there is a Steiner system with $t > 5$.*

We can define more parameters for a t-(v, k, λ) design.

Theorem 10.2. *(Double parameter theorem). Suppose (S, C) is a t-(v, k, λ) design, and let i and j be nonnegative integers satisfying $i + j \leq t$. Then there exists a constant λ_{ij} such that the number of blocks which contain all the elements of any fixed i-set of S and omit all the elements of any fixed j-set of S is exactly λ_{ij}. Furthermore, λ_{ij} is given by*

$$\lambda_{ij} = \frac{\lambda}{\binom{v-t}{k-t}} \binom{v-i-j}{k-i}.$$

Proof. From the inclusion–exclusion principle and Theorem 10.1, we obtain

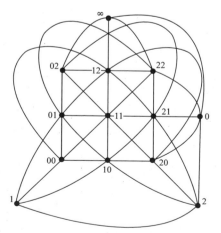

Figure 10.1. A projective plane of order 3.

$$\lambda_{ij} = \sum_{s=0}^{j} \lambda(-1)^s \frac{\binom{v-(i+s)}{k-(i+s)}}{\binom{v-t}{k-t}} \binom{j}{s}$$

$$= \frac{\lambda}{\binom{v-t}{k-t}} \sum_{s=0}^{j} \binom{v-i-s}{k-i-s}(-1)^s \binom{j}{s}.$$

From (1.5) and Exercise 1.15, it follows that

$$\lambda_{ij} = \frac{\lambda}{\binom{v-t}{k-t}} \sum_{s=0}^{j}(-1)^{k-i}\binom{-v+k-1}{k-i-s}\binom{j}{s}.$$

$$= \frac{\lambda}{\binom{v-t}{k-t}}(-1)^{k-i}\binom{-v+k-1+j}{-v+k-1-i+v+1}$$

$$= \frac{\lambda}{\binom{v-t}{k-t}}\binom{-v+k-1+j}{k-i}(-1)^{k-i}$$

$$= \frac{\lambda}{\binom{v-t}{k-t}}\binom{v-i-j}{k-i}.$$

Example. Recall that the Fano configuration *FC* is a 2-(7, 3, 1) design. Applying the double parameter theorem, we determine the constants $\lambda_{00} = 7$, $\lambda_{10} = 3$, $\lambda_{01} = 4$, $\lambda_{20} = 1$, $\lambda_{11} = 2$, $\lambda_{02} = 2$. For example, the relation $\lambda_{11} = 2$ says that given any points x and y of *FC*, there are precisely two lines which contain x and omit y. This can be verified from Figure 9.3.

10.2. BLOCK DESIGNS

A *balanced incomplete block design* (BIBD) is a nontrivial 2-(v, k, λ) design. The parameter theorem (Theorem 10.1) tells us that there is a number r such that each element of the underlying set S occurs in exactly r blocks. Letting $v = |S|$, we rephrase the definition of balanced incomplete block design. A (v, b, r, k, λ) BIBD is a family C of b subsets (*blocks* or *lines* or *edges*) of a set S of v elements (*points* or *vertices*) such that

1. Each point of S lies in exactly r blocks.
2. Each block has k points.

3. Each pair of points of S occur together in λ blocks (the "balance" condition).

4. Not every k-set of S is a block (the "incompleteness" condition).

The nontriviality condition becomes $2 < k < v$.

Examples. FC is a $(7, 7, 3, 3, 1)$ BIBD. FC' is a $(7, 7, 4, 4, 2)$ BIBD. The $S(2, 3, 9)$ Steiner system is a $(9, 12, 4, 3, 1)$ BIBD.

Theorem   10.3. *In   a (v, b, r, k, λ) BIBD, $bk = vr$ and* $r(k - 1) = \lambda(v - 1)$.

Proof.   These equations follow from formulas (10.2) and (10.3) upon letting $t = 2$. The second result is immediate from (10.3). The first result is obtained by dividing equation (10.2) by equation (10.3). $\square$

The equations of Theorem 10.3 can be proved without using the parameter theorem. To prove the first relation, note that the total number of incidences between vertices and blocks is bk (from the point of view of the blocks) and is also vr (from the point of view of the vertices). Therefore, $bk = vr$. To prove the second relation, note that the number of times a particular element occurs in pairs with other elements is $r(k - 1)$. But it is also $\lambda(v - 1)$, the number of elements with which the particular element may be paired multiplied by the number of times each pair occurs. Therefore, $r(k - 1) = \lambda(v - 1)$. For these proofs we have used the technique of enumerating a set in two different ways and equating the results (a technique used in Theorems 8.1 and 10.1).

The incidence relation between blocks and vertices can be displayed in an *incidence matrix* $A = [a_{ij}]$. We choose orderings of the points and the blocks and let $a_{ij} = 1$ if the jth point is an element of the ith block, and $a_{ij} = 0$ otherwise. For instance, the points $1, \ldots, 7$ and the blocks 246, 167, 145, 257, 123, 347, and 356 of FC are represented by the incidence matrix

$$A = \begin{bmatrix} 0 & 1 & 0 & 1 & 0 & 1 & 0 \\ 1 & 0 & 0 & 0 & 0 & 1 & 1 \\ 1 & 0 & 0 & 1 & 1 & 0 & 0 \\ 0 & 1 & 0 & 0 & 1 & 0 & 1 \\ 1 & 1 & 1 & 0 & 0 & 0 & 0 \\ 0 & 0 & 1 & 1 & 0 & 0 & 1 \\ 0 & 0 & 1 & 0 & 1 & 1 & 0 \end{bmatrix}.$$

Theorem 10.4. *For any* (v, b, r, k, λ) *BIBD,* $\det A^t A = rk(r - \lambda)^{v-1}$.

Proof. From the block size and balance condition of a BIBD, it follows that

$$
\det A^t A = \begin{vmatrix} r & \lambda & \cdots & \lambda \\ \lambda & r & & \lambda \\ \vdots & & & \vdots \\ \lambda & \cdots & \lambda & r \end{vmatrix}.
$$

Subtracting the first row from the others, then adding the first column to the other columns we obtain

$$
\det A^t A = \begin{vmatrix} r + \lambda(v - 1) & \lambda & \cdots & \lambda \\ 0 & r - \lambda & & 0 \\ \vdots & & & \vdots \\ 0 & \cdots & 0 & r - \lambda \end{vmatrix} \tag{10.4}
$$

$$
= [r + (v - 1)\lambda](r - \lambda)^{v-1} \tag{10.5}
$$

$$
= rk(r - \lambda)^{v-1}. \tag{10.6}
$$

$\square$

Theorem 10.5. *(Fisher's inequality).* *In any BIBD,* $b \geq v$.

Proof. Let A be an incidence matrix of the (v, b, r, k, λ) BIBD. Thus A is a matrix of dimensions $b \times v$. From the previous theorem it follows that $A^t A = \lambda J_v + (r - \lambda) I_v$ and $\det A^t A = rk(r - \lambda)^{v-1}$. The equation $r(k - 1) = \lambda(v - 1)$ implies $r > \lambda$, which means that $\det(A^t A)$ is not 0. Thus, $v = r(A^t A) \leq \min\{v, b\} \leq b$. $\square$

The extreme case $v = b$ gives rise to an interesting subclass of block designs. A (v, k, λ) *square design* (SD) is a (v, b, r, k, λ) BIBD in which $v = b$ (and hence also $k = r$). Square designs are usually called 'symmetric block designs', but this is probably not the best term for them, as their incidence matrices are not symmetric. We believe that the term 'square design' is a better choice.

Example. FC is a $(7, 3, 1)$ SD and FC' is a $(7, 4, 2)$ SD.

Note that if A is the incidence matrix of an SD, then $(\det A)^2 = k^2(k - \lambda)^{v-1}$, and so $\det A = k(k - \lambda)^{(v-1)/2}$. It follows that if v is even, then $k - \lambda$ is a perfect square. This is the first part of the famous Bruck–Chowla–Ryser theorem, stated below. For a proof of the second part, see Hall (1986).

Theorem 10.6. *(The Bruck–Chowla–Ryser theorem). Suppose that a (v, k, λ) SD exists. Then the following two statements hold:*

(1) If v is even, then $k - \lambda$ is a perfect square.

(2) If v is odd, then the equation

$$x^2 = (k - \lambda)y^2 + (-1)^{(v-1)/2}\lambda z^2$$

has a solution in integers x, y, and z, not all zero.

Example.   There is no $(22, 7, 2)$ SD, for $k - \lambda = 5$, which is not a perfect square.

Theorem 10.7.   *The incidence matrix A of a (v, k, λ) SD is normal: $A^t A = A A^t$. Thus any two distinct blocks intersect in exactly λ elements.*

Proof.   From the proof of Theorem 10.5, $A^t A = (k - \lambda)I + \lambda J$. Because $\det(A^t A)$ is not 0, $\det A$ is not 0, so A^{-1} exists. Therefore $A A^t = A A^t A A^{-1} = A(k - \lambda)IA^{-1} + A\lambda JA^{-1} = (k - \lambda)I + \lambda A J A^{-1}$. Because $JA = kJ$ implies $JA^{-1} = k^{-1}J$, it follows that $A A^t = (k - \lambda)I + \lambda J$. An interpretation of this matrix product yields the desired intersection property. $\square$

The *complement* D' of a BIBD is the design D obtained by switching 0 and 1 in an incidence matrix of D. It is easy to check (Exercise 10.5) that the complement of a (v, b, r, k, λ) BIBD is a $(v, b, b - r, v - k, b - 2r + \lambda)$ BIBD. Specifically, the complement of a (v, k, λ) SD is a $(v, v - k, v - 2k + \lambda)$ SD.

Examples.   The $S(2, 3, 9)$ Steiner system is a $(9, 12, 4, 3, 1)$ BIBD whose complement is a $(9, 12, 8, 6, 5)$ BIBD.

If p is congruent to 3 modulo 4, we can construct a $(p, (p+1)/2, (p+1)/4)$ square design via the set QR of quadratic residues modulo p (see Section 1.5). Recall that if p is any prime greater than 2, then the map $f : \mathbb{Z}_p^* \longrightarrow QR, f(x) = x^2$ is an epimorphism with kernel $\{-1, 1\}$, from which it follows by the first homomorphism theorem for groups that $|QR| = (p-1)/2$. Let QN be the set of quadratic nonresidues modulo p, so that $|QN| = (p-1)/2$. We recall that the *Legendre symbol* (x/p) is defined as follows:

$$\left(\frac{x}{p}\right) = 0 \text{ if } x = 0 \text{ modulo } p$$
$$= 1 \text{ if } x \in QR$$
$$= -1 \text{ if } x \in QN.$$

Because $|QR| = |QN|$, it follows that $\sum(x/p) = 0$ for any sum over a complete residue system modulo p.

Assuming that p is congruent to 3 modulo 4, let A be the $p \times p$ circulant binary matrix whose first row is the characteristic vector of $QR \cup \{0\}$, and whose other rows are successive one unit shifts to the right of the first row. We claim that A is the incidence matrix of a $(p, (p+1)/2, (p+1)/4)$ SD. Evidently, $v = b = p$ and $k = r = (p+1)/2$. We only need to check that the dot product of any two distinct rows is $(p+1)/4$. The dot product of two rows which differ by a shift of k units to the right is

$$\lambda = \frac{1}{2}\left[\left(\frac{k}{p}\right) + 1\right] + \frac{1}{2}\left[\left(\frac{-k}{p}\right) + 1\right] + \frac{1}{4}\sum_{x \in S}\left[\left(\frac{x}{p}\right) + 1\right]\left[\left(\frac{x+k}{p}\right) + 1\right],$$

where S is a complete residue system modulo p except for the values 0 and $-k$. Because p is congruent to 3 modulo 4, -1 is a quadratic nonresidue modulo p. Therefore, exactly one of k and $-k$ is a quadratic residue and the other is a quadratic nonresidue. Therefore, letting x' be the multiplicative inverse of x,

$$\lambda = 1 + \frac{1}{4}\sum_{x \in S}\left[\left(\frac{x}{p}\right)\left(\frac{x+k}{p}\right) + \left(\frac{x+k}{p}\right) + \left(\frac{x}{p}\right) + 1\right]$$

$$= 1 + \frac{1}{4}\sum_{x \in S}\left(\frac{x^2}{p}\right)\left(\frac{1+kx'}{p}\right) + \frac{1}{4}\sum_{x \in S}\left(\frac{x+k}{p}\right) + \frac{1}{4}\sum_{x \in S}\left(\frac{x}{p}\right) + \frac{1}{4}\sum_{x \in S}1$$

$$= 1 + \frac{1}{4}\sum_{x \in S}\left(\frac{1+kx'}{p}\right) - \frac{1}{4}\left(\frac{k}{p}\right) - \frac{1}{4}\left(\frac{-k}{p}\right) + \frac{1}{4}(p-2)$$

$$= \frac{p+2}{4} + \frac{1}{4}\sum_{x \in S}\left(\frac{1+kx'}{p}\right).$$

Because x' takes all values except 0 and $-k'$, $1 + kx'$ takes all values except 1 and 0. Thus,

$$\sum_{x \in S}\left(\frac{1+kx'}{p}\right) = -1,$$

and

$$\lambda = \frac{p+2}{4} - \frac{1}{4}$$
$$= \frac{p+1}{4},$$

as we needed to show.

This construction produces square designs with large values of λ. Such designs are equivalent to the Hadamard designs to be considered in Section 10.6. At the opposite extreme, the next section deals with $\lambda = 1$ designs, which are called projective planes.

10.3.   PROJECTIVE PLANES

A *finite projective plane* π_n of order n is a (v, k, λ) SD in which $\lambda = 1$, $k = n + 1$, $v = n^2 + n + 1$, and $n \geq 2$. The elements of π_n are usually called *points* and the blocks are called *lines*. As a square design, a finite projective plane of order n has the following properties:

1. There are $n^2 + n + 1$ points.
2. There are $n^2 + n + 1$ lines.
3. Every line is incident with $n + 1$ points.

4. Every point is incident with $n + 1$ lines.

5. Every two points determine a unique line.

6. Each pair of lines determines a unique point.

As we mentioned in Section 9.4, the above properties occur in pairs called *duals*. If the words 'point' and 'line' are interchanged, each property is transformed into its dual property.

Example. *FC* is a projective plane of order 2.

Theorem 10.8. *A projective plane π_n exists for any $n = p^k$, where p is prime.*

Proof. Let $F = GF(n)$, the Galois field of order n. Let the set of points of the plane be $S = \{(i,j) : i,j \in F\} \cup \{i : i \in F\} \cup \{\infty\}$. The points $i \in F$ are called *ideal points*. The point ∞ is called the *point at infinity*. The lines are

$$L_{m,b} = \{(x,y) \in F^2 : y = mx + b\} \cup \{m\}, \ 1 \le m, b \le n,$$

$$L_k = \{(k,y) : y \in F\} \cup \{\infty\}, \ 1 \le k \le n, \ and$$

$$L_\infty = \{m : m \in F\} \cup \{\infty\}.$$

(The line L_∞ is called the *line at infinity*.) These $n^2 + n + 1$ points and $n^2 + n + 1$ lines constitute a π_n. We need only check that the conditions for a $(n^2 + n + 1, n^2 + n + 1, n + 1, n + 1, 1)$ BIBD are satisfied. Each point (i,j) lies on exactly $n + 1$ lines, namely, the lines $L_{m,j-mi}$, $1 \le m \le n$, and L_i. The reader should check that the ideal points and the point at infinity each lie on $n + 1$ lines. That each line contains $n + 1$ points can be seen from the definitions. We leave it to the reader to check that each pair of points determines exactly one line. $\square$

Let us use the method in the above proof to construct a projective plane of order 3. The appropriate base field is $F = GF(3)$, whose elements are 0, 1, 2. The 13 points are (i,j), $0 \le i,j \le 2$, which constitute the square array of Figure 10.1, and the ideal points i, $0 \le i \le 2$, and ∞, which are placed to the side. The lines are $L_{m,b}$, $0 \le m, b \le 2$, L_k, $0 \le k \le 2$, and L_∞.

By the Bruck–Chowla–Ryser theorem, if there is a projective plane of order n, and if $n \equiv 1, 2 \bmod 4$, then there is a solution in integers to the equation

$$x^2 = ny^2 - z^2.$$

It follows that

$$(x/y)^2 + (z/y)^2 = n,$$

i.e., n is expressible as the sum of the squares of two rational numbers. From elementary number theory, it follows that n is the sum of two squares of integers. Thus there is no projective plane of order 6 or 14. However, $10 = 3^2 + 1^2$, so the Bruck–Chowla–Ryser theorem does not rule out the possibility of a projective plane of order 10. In 1988 C. Lam, S. Swiercz, L. H. Thiel, and J. McKay used a CRAY-I supercomputer to prove the nonexistence of a projective plane of order 10. See C. Lam, "The search for a finite projective plane of order 10," *American Mathematical Monthly*, **98** (1991), 305–318.

As we showed, there exists a projective plane of every prime power order. No projective plane of order not a prime power is known to exist, and it is conjectured that there is none. It is not known whether there is a projective plane of order 12.

Open Problem 19. *Construct a projective plane of order 12 or show that none exists.*

A *finite affine plane* π'_n is a projective plane of order n without the ideal points and the line at infinity. An affine plane of order n is an $(n^2, n^2 + n, n + 1, n, 1)$ BIBD. For example, a π'_3 is a $(9, 12, 4, 3, 1)$ BIBD, which we have already seen is an $S(2, 3, 9)$ Steiner system. In general, a π'_n is an $S(2, n, n^2)$. Geometrically speaking, these projective and affine planes are finite models of the real Euclidean and projective planes satisfying similar axioms. See Lyndon (1986).

10.4. LATIN SQUARES

A *Latin square* L of order n is an $n \times n$ array $[L(i,j)]$ in which each row and each column contains all of the elements of N_n. An $r \times n$ *Latin rectangle* consists of the first r rows of a Latin square of order n.

Example. The Cayley table of a finite group G with elements $g_1, \ldots, g_n$ yields a Latin square L of order n. Let the (i,j) entry of L be k, where $g_i g_j = g_k$. For instance, $G = \mathcal{Z}_2 \times \mathcal{Z}_2$, with elements $(0,0)$, $(0,1)$, $(1,0)$, $(1,1)$, yields the Latin square

$$\begin{array}{cccc} 1 & 2 & 3 & 4 \\ 2 & 1 & 4 & 3 \\ 3 & 4 & 1 & 2 \\ 4 & 3 & 2 & 1 \end{array}$$

Not all Latin squares come from groups. However, Latin squares are equivalent to the multiplication tables of primitive algebraic structures called quasigroups. We record some interesting definitions. A *groupoid* is a nonempty set S and a binary operation $*$ defined on S. A *semigroup* is a groupoid in which $*$ is an associative operation. A *monoid* is a semigroup containing a two-sided identity element e ($x * e = e * x = x$ for all x). A *group* is a monoid in which every element x in S has a two-sided inverse x^{-1} ($x * x^{-1} = x^{-1} * x = e$). A *quasigroup* is a groupoid such that given $a, b \in S$ there exist unique x, y with $a * x = b$ and $y * a = b$. A *loop* is a quasigroup containing a two-sided identity e. The literature abounds with examples of these algebraic structures. For instance, given any $S(2,3,n)$ Steiner triple system S we may define a quasigroup whose elements are members of S by setting, for a and b distinct, $a * b = c$, where c is the unique element in a triple with a and b, and setting $a * a = a$. By adding an identity element 1 and properly extending the definition of multiplication we may turn this quasigroup into a loop. Another example of a loop is the famous Cayley loop of order 16. To see some other loops the reader should consult Conway and Sloane (1988).

Suppose G is a quasigroup with n elements, $g_1, \ldots, g_n$. By definition, each row and each column of its multiplication table is a permutation of the n elements. Therefore, replacing $g_1, \ldots, g_n$ by the numbers $1, \ldots, n$ results in a Latin square of order n. Conversely, any Latin square of order n is the multiplication table of a quasigroup of order n.

Two Latin squares L_1 and L_2 are *equivalent* if L_1 can be transformed into L_2 by the following operations:

1. Reordering rows.
2. Reordering columns.
3. Permuting symbols.

It is an open problem to determine the number $L(n)$ of Latin squares of order n (equivalently, the number of quasigroups of order n) and the number $L^*(n)$ of inequivalent Latin squares of order n. However, the following existence theorem allows us to formulate a lower bound for $L(n)$.

Theorem 10.9. *(Hall's marriage theorem). Suppose* $S_1, \ldots, S_n$ *are finite sets. There exist distinct* $s_i \in S_i$ *(for each* i*) if the following condition holds for each* k *with* $1 \le k \le n$*: the union of any* k *of the* S_i *contains at least* k *elements.*

The set $\{s_i\}$ is called a *system of distinct representatives* (SDR) for the S_i. If S_i is a list of men whom woman i would like to marry, then an SDR is a feasible set of marriages; hence the title of the theorem.

Proof. Assume that distinct representatives exist for the first k of the sets. We will show that an SDR can be found for $k + 1$ sets. Let T_1 be a set which has no representative assigned to it yet. If there is an element of T_1 not already occurring as a representative of one of the other k sets, then we are done. Otherwise, note that T_1 has at least one element, say t_1, and suppose that t_1 represents T_2. By hypothesis, $T_1 \cup T_2$ contains at least one element other than t_1, say t_2. If t_2 is not already a representative, then stop. If t_2 represents a set T_3, then find $t_3 \in T_1 \cup T_2 \cup T_3$. Continuing in this manner, we find a collection $\{t_i\}$ such that $t_i \in T_1 \cup \cdots \cup T_i$ and t_i represents T_{i+1} (for $i < a$), and t_a is not a representative yet. Now we change some representatives by pairing t_a with a set $T_{a'}$ with $a \le a'$. This process continues until T_1 is paired with a representative. These new pairings, together with the unchanged pairings, constitute an SDR for $k + 1$ sets. $\square$

The following corollary is proved in Hall (1986).

Corollary 10.10. *If* $S_1, \ldots, S_n$ *are sets possessing an SDR, and if the smallest set has size* $t < n$*, then the* S_i *possess at least* $t!$ *SDR's.*

Theorem 10.11. $L(n) \ge n!(n-1)! \ldots 2!1!.$

Proof. We will show that for each r, $1 \le r \le n - 1$, an $r \times n$ Latin rectangle may be extended to an $(r + 1) \times n$ Latin rectangle in at least $(n - r)!$ ways. Given an $r \times n$ Latin rectangle, let S_i be the set of numbers not yet used in column i. Clearly, an SDR could be used as the $(r + 1)$-st row of the Latin square. Now, each element m, $1 \le m \le n$, has occurred in r rows and hence in r columns of the Latin rectangle thus far. Therefore, each element occurs in exactly $n - r$ of the S_i. For each k, the union of k of the S_i contains $k(n - r)$ elements (counting repetitions). As each element occurs in at most $n - r$ of these S_i, the union must contain at least k elements, and the criterion in Hall's theorem is satisfied. Hence, there is an SDR for the S_i.

Because each S_i has size $n - r$, Corollary 10.10 guarantees the existence of at least $(n - r)!$ SDR's. The inequality on $L(n)$ is established by applying the above estimate as each successive row is added to the Latin square. $\square$

By a permutation of its rows and columns, any Latin square may be written with $1, \ldots, n$ as its first row and first column. Such a Latin square is said to be *standardized*. If $L'(n)$ is the number of inequivalent standardized Latin squares of order n, then clearly $L(n) = n!(n - 1)!L'(n)$. Table 10.1 gives all known values of $L'(n)$.

Table 10.1 The value of $L'(n)$ for $1 \leq n \leq 7$.

n	1	2	3	4	5	6	7
$L'(n)$	1	1	1	4	56	9408	16 942 080

Open Problem 20. *Find a formula for $L'(n)$.*

10.5. MOLS AND OODs

Two Latin squares $L_1 = [L_1(i,j)]$ and $L_2 = [L_2(i,j)]$ of order n are *orthogonal* if, for every $(a, b) \in \mathcal{N}_n \times \mathcal{N}_n$, there is an (i, j) with $(L_1(i,j), L_2(i,j)) = (a, b)$. In other words, the ordered pairs $(L_1(i,j), L_2(i,j))$ take each of the n^2 values in $\mathcal{N}_n \times \mathcal{N}_n$ exactly once. The two Latin squares of Figure 10.2 are orthogonal.

A set of *mutually orthogonal Latin squares* or *MOLS* is a set in which every pair is orthogonal. MOLS are also called *pairwise orthogonal Latin squares*. We define $m(n)$ to be the maximum possible number of MOLS of order n. L. Euler introduced the ideas of Latin squares and MOLS in 1782 when he asked whether there are two MOLS of order 6. He believed the answer is no and therefore conjectured that $m(6) = 1$. This was proved by G. Tarry in 1900. Euler also conjectured that $m(n) = 1$ whenever $n \equiv 2 \bmod 4$, but it was shown in 1960 by R. C.

$$1 \quad 2 \quad 3 \qquad\qquad 1 \quad 2 \quad 3$$

$$3 \quad 1 \quad 2 \qquad\qquad 2 \quad 3 \quad 1$$

$$2 \quad 3 \quad 1 \qquad\qquad 3 \quad 2 \quad 1$$

Figure 10.2. Two orthogonal Latin squares.

Bose, E. T. Parker, and S. S. Shrikhande that $m(n) \geq 2$ except when $n = 1$, 2, or 6. For example, this shows that there are two MOLS of order 10. However, it is not known whether there are three MOLS of order 10.

Theorem 10.12. *For all $n \geq 2$, $m(n) \leq n - 1$.*

Proof. Suppose there is a set of n MOLS of order n. By a permutation of symbols, the first row of each Latin square can be changed to $1, \ldots, n$, and permuting symbols clearly does not disturb orthogonality. Now, by the pigeonhole principle, since none of the $(2, 1)$ entries of the n MOLS can equal 1, some two Latin squares have $(2, 1)$ entry equal to i, $2 \leq i \leq n$. But these Latin squares are not orthogonal, because the ordered pair (i, i) occurs twice in the list of ordered pairs of entries. $\square$

Whenever $n = p^k$ for a prime p, we can construct $n - 1$ MOLS of order n. Suppose $F = GF(p^k)$, $F = \{0 = f_0, \ldots, f_{n-1}\}$. For each m, $1 \leq m \leq n - 1$, define the Latin square $L_m = [L_m(i,j)]_{m \times m}$, $0 \leq i, j \leq n - 1$, by $L_m(i,j) = f_m f_i + f_j$. It is a simple matter to check that each L_m is a Latin square. To check the orthogonality condition, observe that $(f_m f_i + f_j, f_n f_i + f_j) = (f_m f_k + f_l, f_n f_k + f_l)$ implies $f_i = f_k$ and $f_j = f_l$, so that all ordered pairs are distinct.

We described in Section 10.3 how a projective plane of order n can be constructed from the field $GF(n)$. The above argument shows that $n - 1$ MOLS can be constructed from $GF(n)$. In fact, a projective plane of order n is equivalent to a set of $n - 1$ MOLS.

Theorem 10.13. *A collection of $n - 1$ MOLS is equivalent to a projective plane of order n.*

Proof. Suppose we are given $n - 1$ MOLS: $L_1, \ldots, L_{n-1}$. Let $S = \{(i,j) \in \mathcal{N}_n \times \mathcal{N}_n\} \cup \{i : i \in \mathcal{N}_{n-1}\} \cup \{0, \infty\}$, and define

$L_{\infty,k} = \{(k,i) : i \in \mathcal{N}_n\} \cup \{\infty\}, 1 \leq k \leq n,$
$L_{0,k} = \{(i,k) : i \in \mathcal{N}_n\} \cup \{0\}, 1 \leq k \leq n,$
$L_{x,y} = \{(i,j) : L_x(i,j) = y\} \cup \{x\}, 1 \leq x \leq n-1, 1 \leq y \leq n,$ and
$L_\infty = \{i : i \in \mathcal{N}_{n-1}\} \cup \{0,\infty\}.$

We leave it to the reader to check that the incidence matrix A for the set of points S and the lines $L_{\infty,k}$, $L_{0,k}$, $L_{x,y}$, L_∞ is an $(n^2 + n + 1, n + 1, 1)$ SD.

Reversing the above construction completes the equivalence. $\square$

The reader may find it profitable to apply the construction technique of Theorem 10.13 to the two MOLS of Figure 10.2 to construct a projective plane of order 3. The plane will be equivalent to the plane of Figure 10.1.

Let us extend the definition of orthogonality to any two square matrices (not necessarily Latin squares). We say that two $n \times n$ matrices A and B with entries from N_n are orthogonal if the ordered pairs $(A(i,j), B(i,j))$ take each of the n^2 values in $\mathcal{N}_n \times \mathcal{N}_n$ exactly once. Notice that, in particular, the matrices

$$R = \begin{bmatrix} 1 & 2 & 3 & \ldots & n \\ 1 & 2 & 3 & \ldots & n \\ & \cdot & & & \\ & \cdot & & & \\ & \cdot & & & \\ 1 & 2 & 3 & \ldots & n \end{bmatrix} \quad \text{and} \quad C = \begin{bmatrix} 1 & 1 & 1 & \ldots & 1 \\ 2 & 2 & 2 & \ldots & 2 \\ & \cdot & & & \\ & \cdot & & & \\ & \cdot & & & \\ n & n & n & \ldots & n \end{bmatrix}$$

are orthogonal. With this generalized definition of orthogonality, we can give an elegant characterization of Latin squares: a matrix L is a Latin square if and only if L is orthogonal to both R and C. Furthermore, it follows that any k MOLS of order n are part of a family of $k + 2$ mutually orthogonal matrices: the k MOLS, R, and C. Conversely, any $k + 2$ mutually orthogonal matrices may be transformed into a set containing R, C, and k MOLS. For if a matrix M is orthogonal to another matrix, then M contains each of the numbers $1, \ldots, n$ exactly n times. Therefore, choosing two matrices M and N from the set of $k + 2$ orthogonal matrices, we may transform M into R and N into C by a simultaneous permutation of the entries of all the matrices. Discarding R and C, we are left with k mutually orthogonal Latin squares.

What all of this accomplishes is the replacement of the notion of Latinicity with the more fundamental notion of orthogonality. Accordingly, we define an *ordered orthogonal design* of order n and depth s (an (n,s) OOD) to be an $s \times n^2$ matrix (M_{ij}) with entries $1, \ldots, n$ such that

$$
\begin{array}{ccccccccc}
1 & 1 & 1 & 2 & 2 & 2 & 3 & 3 & 3 \\
1 & 2 & 3 & 1 & 2 & 3 & 1 & 2 & 3 \\
1 & 2 & 3 & 3 & 1 & 2 & 2 & 3 & 1 \\
1 & 2 & 3 & 2 & 3 & 1 & 3 & 1 & 2 \\
\end{array}
$$

Figure 10.3. A $(3,4)$ OOD.

every two rows are orthogonal. That is, for every pair or rows u and v, every ordered pair (a, b) with $1 \leq a, b \leq n$ occurs exactly once among the ordered pairs (M_{ui}, M_{vi}). For example, Figure 10.3 shows a $(3,4)$ OOD derived from the two Latin squares of order 3 in Figure 10.2. An (n, s) OOD is also referred to as an (n, s) OA (orthogonal array).

The design inherent in Theorem 10.13 can be produced directly from an $(n, n + 1)$ OOD. In general, an *OOD-net* based on an (n, s) OOD is a collection of m points corresponding to the columns of the OOD and s *pencils* of parallel lines, where point x_j is on line y of the ith pencil if the (i, j) entry of the matrix is y. The OOD-net of an $(n, n + 1)$ OOD is an affine plane of order n which may be extended to a projective plane of order n by the addition of one ideal point for each row of the OOD and an ideal line.

In summary, we have shown the equivalence of the following combinatorial configurations:

- a projective plane of order n;
- a collection of $n - 1$ MOLS;
- an $(n, n + 1)$ OOD.

Open Problem 21. *Determine the values of n for which these structures exist.*

A possible conjecture consistent with what is known is that these structures exist if and only if n is a prime power. See Section 10.3.

10.6. HADAMARD MATRICES

In 1893 J. Hadamard considered a basic problem about the maximum absolute value of the determinant of a matrix with bounded entries.

Theorem 10.14. *(Hadamard, 1893). Suppose* $A = [a_{ij}]$ *is matrix of order* n *with* $-1 \leq a_{ij} \leq 1$ *for all* i *and* j. *An upper bound for* $|\det A|$ *is* $n^{n/2}$, *and this value is achieved if and only if* $a_{ij} = \pm 1$ *for all* i *and* j *and* $AA^t = nI$.

Proof. The rows of A are vectors in $\mathcal{R}^n$ of length at most $n^{1/2}$, and they span a parallelepiped of volume $|\det A|$. This volume is clearly maximized when the vectors are mutually orthogonal and of maximum possible length, and in this case the volume is the product of the lengths, $n^{n/2}$. $\sqcup$

A matrix $A = [a_{ij}]$ with $a_{ij} = \pm 1$ and $AA^t = nI$ is called a *Hadamard matrix* of order n. The condition $AA^t = nI$ means that the dot product of any two distinct rows of A is zero. (The same is true for columns, as $A^t A = A^{-1}(AA^t)A = A^{-1}(nI)A = nI$.) Therefore, without regard to the volume argument given in the proof above, if A is a Hadamard matrix of order n, then $n^n = \det AA^t = (\det A)^2$, which implies that $|\det A| = n^{n/2}$.

Notice that Theorem 10.14 does not address the question of the maximum value of $|\det A|$ when there is no Hadamard matrix of order n. However, $|\det A|$ does attain a maximum, as it is a continuous function defined on a compact set (the cube $[-1, 1]^{n^2}$). Furthermore, because $\det A$ is a linear function of each entry a_{ij} (i.e., a straight line), the function $y = |\det A|$ is concave upward and therefore the maximum of $|\det A|$ occurs when each $a_{ij} = -1$ or 1.

The *Kronecker product* $A \otimes B$ of two square matrices $A = [a_{ij}]_{m \times m}$ and $B = [b_{ij}]_{n \times n}$ is the square matrix $A \otimes B = [a_{ij}B]_{mn \times mn}$. The Kronecker product produces larger Hadamard matrices from smaller ones. For example, Figure 10.4 shows Hadamard matrices A and B of orders 2 and 4, respectively, with $B = A \otimes A$.

$$A = \begin{bmatrix} 1 & 1 \\ -1 & 1 \end{bmatrix} \qquad B = \begin{bmatrix} 1 & 1 & 1 & 1 \\ -1 & 1 & -1 & 1 \\ -1 & -1 & 1 & 1 \\ 1 & -1 & -1 & 1 \end{bmatrix}$$

Figure 10.4. Hadamard matrices of orders 2 and 4.

In general, if H is a Hadamard matrix, then so is

$$\begin{bmatrix} H & H \\ -H & H \end{bmatrix}.$$

Theorem 10.15. *For any $n = 2^m$, where m is a positive integer, there is a Hadamard matrix of order n.*

If A is a Hadamard matrix, then any permutation of the rows or columns of A is a Hadamard matrix. Also, any row or column of A may be multiplied by -1 with the result still a Hadamard matrix. With these operations it is possible to alter any Hadamard matrix so that its first row and first column consist of all 1's. Such a Hadamard matrix is said to be *normalized*.

Theorem 10.16. *If A is a Hadamard matrix of order $n > 2$, then n is a multiple of 4.*

Proof. Normalize A and permute its columns so that its first three rows look like this:

$$
\begin{array}{cccccccccccccccc}
1 & 1 & 1 & \ldots & 1 & 1 & 1 & \ldots & 1 & 1 & 1 & \ldots & 1 & 1 & 1 & \ldots \\
1 & 1 & 1 & \ldots & 1 & 1 & 1 & \ldots & -1 & -1 & -1 & \ldots & -1 & -1 & -1 & \ldots \\
1 & 1 & 1 & \ldots & -1 & -1 & -1 & \ldots & 1 & 1 & 1 & \ldots & -1 & -1 & -1 & \ldots \\
& a & & & & b & & & & c & & & & d & &
\end{array}
$$

(The variables a, b, c, d are to be determined.) Four equations are immediate from the definition of a Hadamard matrix:

$$a + b + c + d = n$$
$$a + b - c - d = 0$$
$$a - b + c - d = 0$$
$$a - b - c + d = 0.$$

Adding the equations yields $4a = n$, from which it follows that $a = b = c = d = n/4$. Therefore, n is a multiple of 4, and furthermore we have found that any row after the first has $(n/2)$ $+1$'s and $(n/2)$ -1's, and any two rows (not including the first) have $+1$'s together in $n/4$ columns. $\square$

The smallest order for which the existence of a Hadamard matrix is not certain is 428.

Open Problem 22. *Determine whether there is a Hadamard matrix of order 428.*

Theorem 10.17. *Let H be a normalized Hadamard matrix of order $n = 4m \geq 8$. Deleting the first row and column of H and changing each -1 to a 0 results in the incidence matrix of a $(4m - 1, 2m - 1, m - 1)$ SD.*

Proof. Let X be the submatrix of H formed by deleting its first row and column. The Hadamard conditions imply that $XJ = JX = -J$ and $XX^t = 4mI - J$. When each -1 is switched to 0 a new matrix $Y = \frac{1}{2}(X + J)$ results. We check that Y is the incidence matrix of a $(4m - 1, 2m - 1, m - 1)$ SD: $JY = JY = \frac{1}{2}(-J + (4m - 1)J) = (2m - 1)J$ and $YY^t = \frac{1}{4}(X + J)(X^t + J) = mI + (m - 1)J$. $\square$

A $(4m - 1, 2m - 1, m - 1)$ SD created this way is called a *Hadamard design*. A 3-$(4m, 2m, m - 1)$ design may be formed by taking complements of a Hadamard design H together with the blocks of H joined to a new point ∞.

Hadamard designs and projective planes are the two extremal types of (v, k, λ) designs. For if a (v, k, λ) design exists, then $4n - 1 \leq v \leq n^2 + n + 1$, where $n = k - \lambda$. To prove this inequality, let $\lambda' = v - 2k + \lambda$. Then $\lambda + \lambda' = v - 2n$ and $\lambda\lambda' = n(n - 1)$. The upper bound follows from the observation that $(\lambda + \lambda')^2 \geq 4\lambda\lambda'$, and the lower bound follows from the inequality $\lambda \geq 1$. One can go on to show that the upper bound is met if and only if the design is a projective plane (or its complement) and the lower bound is met only for a Hadamard design (or its complement).

Let H be a normalized Hadamard matrix of order $n = 4m$, with each -1 changed to 0, and let A be the code consisting of the rows of H and the binary complements of these rows. Clearly, A is a code of dimension $4m$ containing $8m$ codewords. We claim that the distance of A is $2m$. The proof of Theorem 10.16 guarantees that any two rows of H disagree in exactly $2m$ places. Therefore, any two rows of H^c disagree in $2m$ places. Suppose $a \in H$ and $b \in H^c$. If $a = b^c$, then $d(a, b) = n$. If not, then $d(a, b)$ equals the number of components in which a and b^c agree, which is $2m$. This $(4m, 8m, 2m)$ code is called a *Hadamard code*. It is capable of detecting $2m - 1$ errors and correcting $m - 1$ errors.

Example. The $(7, 3, 1)$ SD yields an $(8, 16, 4)$ code.

Example. With a little trial and error, or by taking the complement of the quadratic residues construction for $p = 11$ described in Section 10.2, we obtain the incidence matrix of an $(11, 5, 2)$ SD:

$$Y = \begin{bmatrix}
1 & 0 & 1 & 1 & 1 & 0 & 0 & 0 & 1 & 0 & 0 \\
0 & 1 & 0 & 1 & 1 & 1 & 0 & 0 & 0 & 1 & 0 \\
0 & 0 & 1 & 0 & 1 & 1 & 1 & 0 & 0 & 0 & 1 \\
1 & 0 & 0 & 1 & 0 & 1 & 1 & 1 & 0 & 0 & 0 \\
0 & 1 & 0 & 0 & 1 & 0 & 1 & 1 & 1 & 0 & 0 \\
0 & 0 & 1 & 0 & 0 & 1 & 0 & 1 & 1 & 1 & 0 \\
0 & 0 & 0 & 1 & 0 & 0 & 1 & 0 & 1 & 1 & 1 \\
1 & 0 & 0 & 0 & 1 & 0 & 0 & 1 & 0 & 1 & 1 \\
1 & 1 & 0 & 0 & 0 & 1 & 0 & 0 & 1 & 0 & 1 \\
1 & 1 & 1 & 0 & 0 & 0 & 1 & 0 & 0 & 1 & 0 \\
0 & 1 & 1 & 1 & 0 & 0 & 0 & 1 & 0 & 0 & 1
\end{bmatrix}.$$

From Y we construct a normalized Hadamard matrix of order 12. Thus,

$$H = \begin{bmatrix}
1 & 1 & 1 & 1 & 1 & 1 & 1 & 1 & 1 & 1 & 1 & 1 \\
1 & 1 & -1 & 1 & 1 & 1 & -1 & -1 & -1 & 1 & -1 & -1 \\
1 & -1 & 1 & -1 & 1 & 1 & 1 & -1 & -1 & -1 & 1 & -1 \\
1 & -1 & -1 & 1 & -1 & 1 & 1 & 1 & -1 & -1 & -1 & 1 \\
1 & 1 & -1 & -1 & 1 & -1 & 1 & 1 & 1 & -1 & -1 & -1 \\
1 & -1 & 1 & -1 & -1 & 1 & -1 & 1 & 1 & 1 & -1 & -1 \\
1 & -1 & -1 & 1 & -1 & -1 & 1 & -1 & 1 & 1 & 1 & -1 \\
1 & -1 & -1 & -1 & 1 & -1 & -1 & 1 & -1 & 1 & 1 & 1 \\
1 & 1 & -1 & -1 & -1 & 1 & -1 & -1 & 1 & -1 & 1 & 1 \\
1 & 1 & 1 & -1 & -1 & -1 & 1 & -1 & -1 & 1 & -1 & 1 \\
1 & 1 & 1 & 1 & -1 & -1 & -1 & 1 & -1 & -1 & 1 & -1 \\
1 & -1 & 1 & 1 & 1 & -1 & -1 & -1 & 1 & -1 & -1 & 1
\end{bmatrix}.$$

We shall see in the next section that Y is one of the main ingredients in producing a perfect $(23, 2^{12}, 7)$ code. Switching -1 back to 0, the rows of H and their complements constitute a $(12, 24, 6)$ code which detects 5 errors and corrects 2 errors. The words of this code are

```
1  1  1  1  1  1  1  1  1  1  1  1
1  1  0  1  1  1  0  0  0  1  0  0
1  0  1  0  1  1  1  0  0  0  1  0
1  0  0  1  0  1  1  1  0  0  0  1
1  1  0  0  1  0  1  1  1  0  0  0
1  0  1  0  0  1  0  1  1  1  0  0
1  0  0  1  0  0  1  0  1  1  1  0
1  0  0  0  1  0  0  1  0  1  1  1
1  1  0  0  0  1  0  0  1  0  1  1
1  1  1  0  0  0  1  0  0  1  0  1
1  1  1  1  0  0  0  1  0  0  1  0
1  0  1  1  1  0  0  0  1  0  0  1
```

```
0  0  0  0  0  0  0  0  0  0  0  0
0  0  1  0  0  0  1  1  1  0  1  1
0  1  0  1  0  0  0  1  1  1  0  1
0  1  1  0  1  0  0  0  1  1  1  0
0  0  1  1  0  1  0  0  0  1  1  1
0  1  0  1  1  0  1  0  0  0  1  1
0  1  1  0  1  1  0  1  0  0  0  1
0  1  1  1  0  1  1  0  1  0  0  0
0  0  1  1  1  0  1  1  0  1  0  0
0  0  0  1  1  1  0  1  1  0  1  0
0  0  0  0  1  1  1  0  1  1  0  1
0  1  0  0  0  1  1  1  0  1  1  0
```

The families of codes we have encountered can be arranged on a continuum from good rate/bad distance to bad rate/good distance, as in Figure 10.5. At the left extreme is the code F^n, in which every vector is a codeword. The rate is 1, but the code is incapable of correcting any errors. At the right extreme is a code consisting of any vector v and its complement v^c. Although this code has the highest possible distance, $d = n$, its rate is $1/n$, the lowest possible. The family of Hamming codes are capable of correcting $e = 1$ error, and the rates tend to 1. Therefore, the Hamming codes converge to the left endpoint of the continuum. (The triplicate code corrects $e = 1$ error and has rate $\frac{1}{3}$, and should therefore be put beneath the family of Hamming codes, not on the continuum.) For the Hadamard codes,

$$r(A) = \frac{\log_2 8m}{4m} = \frac{3 + \log_2 m}{4m}.$$

F^n Hamming codes   Golay code G_{23} Hadamard codes $\{v, v^c\}$

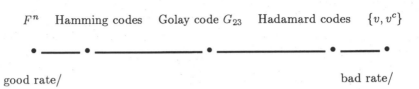

good rate/   bad rate/

bad distance   good distance

Figure 10.5.   The world of codes.

It follows that the rates of these codes converge to 0. Therefore, on the continuum, the Hadamard codes converge to the right endpoint.

The $(23, 2^{12}, 7)$ Golay code G_{23}, which we will construct in the next chapter, has rate $\frac{12}{23}$ and corrects $e = 3$ errors.

Notes

Theorem 10.5 was proved by the statistician and biologist R. A. Fisher (1890–1962).

In general, a projective plane over a field F has automorphism group Aut $F \cdot PGL(3, F)$. If F is $GF(q)$, $q = p^k$, then the order of the group is $k(q^3 - q)(q^3 - q^2)(q^2 + q + 1)$ elements.

Many t-designs are known with $t \le 3$ but not many with $t \ge 4$. However, L. Teirlinck in 1987 proved the existence of t-designs for all values of t.

P. Hall published his marriage theorem in 1935. See Harary (1969) for a description of some equivalent theorems, such as the König–Egerváry theorem and Menger's theorem. The König–Egerváry theorem states that the minimum number of rows and columns which cover the 1's in a 0–1 matrix equals the maximum number of row- and column-independent 1's in the matrix. Menger's theorem states that the minimum number of vertices which separate two given nonadjacent vertices in a finite connected graph equals the maximum number of edge-disjoint paths which connect the two vertices.

Exercises

10.1   Draw the $S(2, 3, 9)$ Steiner system.

10.2   Construct a 2-$(21, 5, 1)$ design.

10.3   Show that in an $S(4, 5, 11)$ Steiner system no two blocks are disjoint. Show that in an $S(5, 6, 12)$ Steiner system the complement of any block is a block. [Hint: Let $\{a, b, c, d, e\}$ be a block in the $S(4, 5, 11)$ Steiner system. Let A be the set of blocks containing a, B the set of blocks containing b, etc. Use the inclusion–exclusion principle and knowledge of the values of λ_i to find $|A \cup B \cup C \cup D \cup E|$. Solve the problem about $S(5, 6, 12)$ similarly.]

10.4 Show that the double parameters λ_{ij} satisfy the relations $\lambda_{i0} = \lambda_i$ and $\lambda_{(i-1)j} = \lambda_{(i-1)(j-1)} - \lambda_{i(j-1)}$. Show that from these relations the values of λ_{ij} for all $i + j \leq t$ can be calculated.

10.5 Prove that the complement of a (v, b, r, k, λ) BIBD is a $(v, b, b - r, v - k, b - 2r + \lambda)$ BIBD.

10.6 Draw a projective plane of order 4.

10.7 Find a circulant incidence matrix for FC.

10.8 Let M be the 13×13 circulant matrix whose first row is the characteristic vector of the set $\{1, 2, 4, 10\}$. Show that M is an incidence matrix for a $(13, 4, 1)$ SD, i.e., a projective plane of order 3.

10.9 (Putnam Competition, 1954) Let n be an odd integer greater than 1. Let A be an n by n symmetric matrix such that each row and each column of A consists of some permutation of the integers $1, \ldots, n$. Show that each one of the integers $1, \ldots, n$ must appear in the main diagonal of A.
[Hint: Because A is symmetric the off-diagonal entries occur in pairs.]

10.10 Construct a system of four 5×5 MOLS.

10.11 Use the Kronecker product to construct a Hadamard matrix of order 8. Use the method of Theorem 10.17 to change this into a $(7, 3, 1)$ SD, i.e., a Fano configuration. Show that the code produced from this matrix is the code with parameters $(8, 16, 4)$ asked for in Exercise 9.4.

10.12 Use the Kronecker product to construct a Hadamard matrix of order 16. Construct a $(15, 7, 3)$ SD. What code does this design give?

10.13 Construct a $(19, 9, 4)$ SD. What code does this design give?

11

Big Designs

We conclude the discussion of combinatorial designs by constructing three large, interesting, related configurations, namely, the $(23, 2^{12}, 7)$ Golay code G_{23} (the only perfect multi-error-correcting binary code), the $S(5, 8, 24)$ Steiner system (consisting of the weight 8 codewords of the extended Golay code G_{24}), and Leech's lattice L (a 24-dimensional lattice obtained from G_{24} which generates a surprisingly tight sphere-packing).

11.1. THE GOLAY CODES AND S(5, 8, 24)

Theorem 9.3 states that the only feasible parameters for perfect binary codes with $e > 1$ are $(90, 2^{78}, 5)$ and $(23, 2^{12}, 7)$. Exercise 9.7 calls for a proof that no $(90, 2^{78}, 5)$ code exists. We now construct a $(23, 2^{12}, 7)$ code called the *Golay code*, G_{23}. It is a perfect 3-error correcting code with 2^{12} words, sitting inside F^{23}. As a bonus, we will find that certain words in the extended Golay code G_{24} constitute a Steiner system $S(5, 8, 24)$. At the end of the previous chapter we constructed an $(11, 5, 2)$ SD. Let M be the complement of this design. It is easy to check (Exercise 10.5) that M is an $(11, 6, 3)$ SD. Let G be the 12×24 matrix

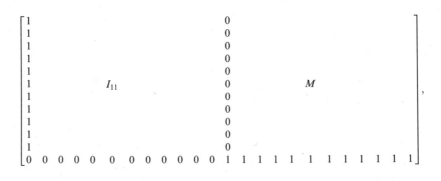

where I_{11} is the order 11 identity matrix.

The matrix G defines a linear transformation $G : F^{12} \longrightarrow F^{24}$, $x \longmapsto xG$. We leave it to the reader to use elementary row operations to reduce G to row echelon form and thereby show that G has rank 12. Thus the code $G_{24} = \{xG : x \in F^{12}\}$ generated by G has parameters $(24, 2^{12}, d)$, where d is to be determined. G is called a *generating matrix* for G_{24}.

We observe that if a and b are any two vectors of the same length, then $w(a + b) = w(a) + w(b) - 2a \cdot b$. This may be proved componentwise.

If a and b are different rows of G, then $a \cdot b$ is even. This may be proven by trying all possible row products.

Now we show that if x is in G_{24}, then $w(x)$ is a multiple of 4. Any codeword is a linear combination (over F) of the rows of G, so we can write $x = r_1 + r_2 + \cdots + r_n$ (with relabeling of rows). We use induction on n. If $n = 1$, then by inspection $w(x)$ is a multiple of 4 (the last row has weight 12 and every other row has weight 8). Now if $x = r_1 + r_2 + \cdots + r_n + r_{n+1}$ then $w(x) = w(r_1 + \cdots + r_n) + w(r_{n+1}) - 2(r_1 + \cdots + r_n) \cdot r_{n+1}$, which, by remarks made earlier, is a multiple of 4. This completes the induction.

Therefore, the possible weights of words of G_{24} are 0, 4, 8, 12, 16, 20, 24. Clearly, $0 \in G_{24}$ and $w(0) = 0$. Also, $r_1 + r_2 + \cdots + r_{24} = (1, 1, \ldots, 1, 1)$, so there is a word of weight 24. It follows that the binary complement of any codeword is also a codeword, and therefore that the weight distribution of G_{24} is symmetric. The weight distribution of G_{24}, as it is known so far, is given by Table 11.1. The variables α, β, γ have yet to be determined.

Suppose $x \in G_{24}$. Let $L(x)$ be the left-hand string of length 12 of x and $R(x)$ the right-hand string. We can now represent x as $x = [L(x), R(x)]$. We claim that $[R(x), L(x)] \in G_{24}$; that is, G_{24} is invariant under the permutation of coordinates

$$\tau = (1\ 13)(2\ 14)(3\ 15) \ldots (12\ 24).$$

We say that G_{24} is *self-symmetric*. Let v' denote the vector obtained from v by switching the right and left halves. We leave it to the reader to show

Table 11.1 The partially known weight distribution of G_{24}.

weight	0	4	8	12	16	20	24
number of words	1	α	β	γ	β	α	1

that for each row of G, r' can be written as a linear combination of the rows r_i. It follows that G_{24} is invariant under τ, as every codeword of G_{24} is a sum of rows r_i.

Next we observe that $w(L(x))$ is even whenever $x \in G_{24}$, because the sum of k rows when k is even yields $w(L(x)) = k$, and the sum when k is odd yields $w(L(x)) = k + 1$.

Because $w(L(x)) + w(R(x))$ is a multiple of 4, it follows that $w(R(x))$ is always even.

Finally, we will show $d \geq 8$, from which it follows that $d = 8$, because the first row of G has weight 8. We need only show that no codeword has weight 4. If x has weight 4, then $(w(L(x)), w(R(x))) = (0, 4), (4, 0)$, or $(2, 2)$. If $w(L(x)) = 0$, then x must be r_{12}; but then $w(R(x)) = 8 > 4$. Because G_{24} is self-symmetric, the $(4, 0)$ case is ruled out also. For the $(2, 2)$ case, we can sum any one or two of the first eleven rows of G and then add or not add the twelfth row. In each case the resulting codeword has weight 6, not 2.

Therefore, G_{24} is a $(24, 2^{12}, 8)$ code. The deletion of any coordinate of G_{24} produces the Golay code G_{23}, a code with parameters $(23, 2^{12}, 7)$.

Theorem 11.1.   *The Golay code G_{23} is a (perfect) $(23, 2^{12}, 7)$ code.*

Let us complete the weight distribution table of G_{24}. We know G_{24} has no codewords of weight 4 or 20. How many words have weight 8? If $w(x) = 8$ then $(w(L(x)), w(R(x)))$ equals $(0, 8), (2, 6), (4, 4), (6, 2)$, or $(8, 0)$. A glance at G shows that $(0, 8)$ is impossible, and hence by self-symmetry $(8, 0)$ is impossible. To obtain a weight partition $(2, 6)$, we can add one or two rows of the first eleven rows of G and then add or not add the twelfth row. There are $2(\binom{11}{1} + \binom{11}{2}) = 132$ possibilities. Likewise, there are 132 ways of arriving at a codeword with weight partition $(6, 2)$. To get a $(4, 4)$ weight partition, we add either three or four rows of G. The number of choices is $\binom{11}{3} + \binom{11}{4} = 495$. Altogether, the number of words of weight 8 in G_{24} is $132 + 132 + 495 = 759$.

We display the complete weight distribution of G_{24} in Table 11.2.

We are now ready to find the Steiner system $S(5, 8, 24)$ sitting inside G_{24}. From the parameter theorem (Theorem 10.1) we obtain the following structural constants for $S(5, 8, 24)$: $\lambda_5 = 1$, $\lambda_4 = 5$, $\lambda_3 = 21$, $\lambda_2 = 77$,

Table 11.2 The weight distribution of G_{24}

weight	0	8	12	16	24
number of words	1	759	2,576	759	1

$\lambda_1 = 253$, $\lambda_0 \doteq 759$. As λ_0 is the number of blocks, it is clear that the words of weight 8 in G_{24} should form the blocks of $S(5, 8, 24)$. Let S be the set of coordinates $1, \ldots, 24$ and let C be the collection of sets of 8 coordinates which equal 1 in codewords of weight 8 in G_{24}. We need to check that the $t = 5$ and $\lambda = 1$ conditions are met. That is, we must see that every 5-element subset of S is contained in exactly one block in C. This is equivalent to every vector of weight 5 being covered by exactly one codeword of weight 8 in G_{24}.

A vector v of weight 5 cannot be covered by two different codewords x and y of weight 8, or else $w(x + y) \leq 6$, a contradiction.

Therefore, it remains to demonstrate that there are enough codewords of weight 8 to satisfy the $\lambda = 1$ condition. There are $\binom{24}{5}$ vectors of weight 5, and each of the 759 codewords of weight 8 covers $\binom{8}{5}$ of them. A simple calculation shows that $\binom{24}{5} = 759\binom{8}{5}$. Therefore, (S, C) is the desired Steiner system $S(5, 8, 24)$.

Theorem 11.2. *The design (S, C) is an $S(5, 8, 24)$ Steiner system.*

11.2. LATTICES AND SPHERE PACKINGS

An *n-dimensional lattice* X is a subset of $\mathcal{R}^n$ such that

1. $0 = (0, \ldots, 0) \in X$.
2. $x \in X$ implies $-x \in X$.
3. $x, y \in X$ imply $x + y \in X$.

We also assume that X contains a point other than the origin (hence X is infinite), and that X is *discrete*, which means that X has a finite intersection with any compact subset of $\mathcal{R}^n$.

The *distance* between two points x and y in $\mathcal{R}^n$ is

$$d(x, y) = ((x_1 - y_1)^2 + \cdots + (x_n - y_n)^2)^{\frac{1}{2}},$$

and the *norm* of x is

$$\|x\| = d(x, 0) = (x_1^2 + \cdots + x_n^2)^{\frac{1}{2}}.$$

If X is a discrete *n*-dimensional lattice containing a nonorigin point x, then the *Euclidean sphere* $S(0, x) = \{y \in \mathcal{R}^n : 0 \leq \|y\| \leq \|x\|\}$ contains only finitely many points of X. The minimum positive distance between

0 and a point in this intersection is the *minimum distance* of X, denoted $b(X)$. Two lattice points are *neighbors* if they are separated by this minimum distance. Because a lattice is clearly translation invariant, any base point would determine the same minimum distance $b(X)$ to a neighbor. If spheres of radius $\frac{1}{2}b(X)$ are centered at each lattice point, these spheres touch only at their surfaces. Such a placement of spheres is called a *lattice sphere packing* of $\mathcal{R}^n$.

It is desirable to know how densely spheres may be packed in $\mathcal{R}^n$, via a lattice packing or otherwise. Related to this question is the computation of the maximum number of spheres touching a given sphere in a lattice packing. Specifically, let $c(X)$, the *contact number* of X, be the number of spheres of radius $\frac{1}{2}b(X)$ which touch the sphere centered at the origin. Equivalently, $c(X)$ is the number of lattice points at distance $b(X)$ from 0.

Figure 11.1 shows part of a 2-dimensional lattice with distance 2 and contact number 4. This lattice is called $(2\mathcal{Z})^2$, as it consists of all points in the plane with two even integer coordinates. In general, the n-dimensional lattice $(2\mathcal{Z})^n$ consists of all points in $\mathcal{R}^n$ with n even integer coordinates.

We can easily calculate the amount of the plane enclosed by the circles. Consider a square of area 4 whose vertices are four lattice points (as indicated in the figure). Such a square is called a *fundamental region* of the packing, as this region repeats periodically to give the complete pattern of the packing. Because the square contains four circular quadrants whose total area is π, we say that the *density* of the packing is $\frac{\pi}{4} \doteq 0.79$. However, the densest possible packing in $\mathcal{R}^2$ is not based on the lattice $(2\mathcal{Z})^2$, but instead is based on the lattice T (*triangular lattice*) shown in Figure 11.2. The fundamental region of T is an equilateral

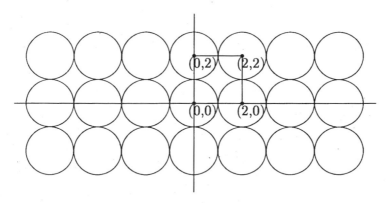

Figure 11.1.   A lattice packing with contact number 4 and density ≈ 0.79.

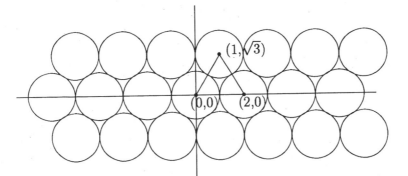

Figure 11.2. A lattice packing with contact number 6 and density ≈ 0.90.

triangle of side length 2 and area $\sqrt{3}$. As this triangle contains three sixths of a unit circle, the density of the packing is $\frac{\pi}{2\sqrt{3}} \doteq 0.90$. The contact number, six, is also greater for T than for $(2\mathcal{Z})^2$. It is generally easier to determine the contact number of a lattice than its density. For example, because a unit sphere in the $(2\mathcal{Z})^n$ packing has two neighbors in each of n "directions", it follows that the lattice $(2\mathcal{Z})^n$ has contact number $2n$. Surprisingly, we shall see that this contact number can be greatly improved upon in 24 dimensions.

Open Problem 23. *Find, for each k, the highest possible contact number of a lattice sphere packing in $\mathcal{R}^k$.*

11.3. LEECH'S LATTICE

In 1965 J. Leech noticed that the Golay code G_{24} can be used to construct a 24-dimensional lattice with remarkably high contact number. Leech's lattice L is defined as follows. For each codeword $c = (c_1, \ldots, c_{24}) \in G_{24}$, and each integer m, let $c[m]$ be the set of integer 24-tuples $(x_1, \ldots, x_{24})$ for which

$$(A) \sum x_i = 4m \text{ and}$$
$$(B) \quad x_i \equiv m \bmod 4 \text{ if } c_i = 0$$
$$\equiv m + 2 \bmod 4 \text{ if } c_i = 1.$$

Leech's lattice L is the union of all the $c[m]$.

We need to verify that L is a lattice. Because the all zero string is a codeword in G_{24}, it follows that $0 \in L$ (with $m = 0$). If $x \in L$, then

$x \in c[m]$ for some $c \in G_{24}$ and some integer m. It is easy to show that $-x \in c[-m]$, and hence $-x \in L$. We will show that if $x, y \in L$ with $x \in c[m]$ and $y \in d[n]$, then $x + y \in (c + d)[m + n]$. Clearly, $\sum(x_i + y_i) = \sum x_i + \sum y_i = 4m + 4n = 4(m + n)$, so that condition (A) is satisfied. As for condition (B), if c_i and d_i are of opposite parity, then $c_i + d_i = 1$ and $x_i + y_i \equiv m + n + 2 \bmod 4$. If c_i and d_i have the same parity, then $c_i + d_i = 0$ and $x_i + y_i \equiv m + n \bmod 4$. Therefore, condition (B) is satisfied and L is a lattice.

We now calculate the distance $b(L)$ of Leech's lattice by finding the smallest value of $\|x\| = (x_1^2 + \cdots + x_{24}^2)^{1/2}$ for a lattice point other than the origin. Condition (B) implies that all the x_i are even (if m is even) or all the x_i are odd (if m is odd). If all the x_i are odd and some x_i satisfies $\|x_i\| \geq 3$, then $\sum x_i^2 \geq 23(1) + 3^2(1) = 32$. We shall soon see that 32 is, in fact, the minimum value of $\|x\|^2$. Recall that the codewords of G_{24} have weights 0, 8, 12, 16, and 24. We say they have *shapes* 0^{24}, $0^{16}1^8$, $0^{12}1^12$, 0^81^16, and 1^{24}. If $\|x_i\| = 1$ for all x_i, then x has shape $(+1)^a(-1)^b$ where (a, b) is one of $(24, 0)$, $(16, 8)$, $(12, 12)$, $(8, 16)$, $(0, 24)$. Therefore, $\sum x_i = 24, 8, 0, -8,$ or -24. In any case, the sum is $4m$ with m even, a contradiction. Hence, the minimum value of $\|x\|$ for odd x_i is $32^{1/2}$ and is achievable only by lattice points of shape $\pm 1^{23} \pm 3$.

Now suppose all the x_i are even. If $\|x_i\| > 4$ for some x_i, then $\sum x_i^2 \geq 6^2 > 32$, so we can disregard these vectors and assume $\|x_i\| = 0, 2,$ or 4. If at least one x_i has $\|x_i\| = 2$, then at least eight do (by examining the shapes of the codewords). Therefore, $\sum x_i^2 \geq 8 \cdot 2^2 = 32$. If always $\|x_i\| = 0$ or 4, then $\|x_i\| = 4$ for at least two x_i, or else $\sum x_i = \pm 4 = 4(\pm 1)$, contradicting the fact that m is even. If $\|x_i\| = 4$ for more than two x_i, then $\|x\|$ is too large. Thus the minimum value of $\|x\|$ for even x_i is $32^{1/2}$ and is achievable only by lattice points of shape $0^{16} \pm 2^8$ or $0^{22} \pm 4^2$.

Theorem 11.3. *The distance of Leech's lattice is $b(L) = 32^{1/2}$, and the neighbors of the origin have shapes $\pm 1^{23} \pm 3$, $0^{16} \pm 2^8$, and $0^{22} \pm 4^2$.*

We now calculate the contact number $c(L)$ of Leech's lattice, noting first that each lattice point comes from a unique choice of c and m. Recalling Table 11.2, there are 759 codewords of weight 8 in G_{24}. Each word gives rise to 2^7 lattice points of shape $0^{16} \pm 2^8$, for the m even condition forces an even number of $+2$ and -2 components. Therefore, the signs of the first seven 2's may be chosen arbitrarily and the sign of the eighth 2 is forced. Thus there are $2^7 \cdot 759$ lattice points of shape $0^{16} \pm 2^8$.

How many lattice points have shape $0^{22} \pm 4^2$? Any choice of signs for the two 4's satisfies the m even condition. Therefore, because there are $\binom{24}{2}$ choices for the placement of the 4's (lattice points of this shape come from the all 0 codeword and the all 1 codeword), there are $2^2 \binom{24}{2}$ lattice points of shape $0^{22} \pm 4^2$.

To find the number of lattice points of shape $\pm 1^{23} \pm 3$, let z be the number of $+1$'s in a lattice vector. Then $4m = z(1) - (23 - z) \pm 3 = 2z - 23 \pm 3$. This equation forces the $+3$ to occur if z is even and the -3 if z is odd. Any of the 2^{12} codewords generates a choice of $+1$'s and -1's. The position of the ± 3 may be chosen in 24 ways, and once it is chosen the sign is determined by the previous comment. Thus there are $2^{12} \cdot 24$ lattice points of shape $\pm 1^{23} \pm 3$.

Therefore, $c(L) = 2^7 \cdot 759 + 2^2 \binom{24}{2} + 2^{12} \cdot 24 = 196\,560$.

Theorem 11.4.   *The contact number of Leech's lattice is $c(L) = 196\,560$. Neighbors of the origin, listed by shape, occur with the following multiplicities:*

Shape	Number
$0^{16} \pm 2^8$	$2^7 \cdot 749 = 97\,152$
$0^{22} \pm 4^2$	$2^2 \binom{24}{2} = 1\,104$
$\pm 1^{23} \pm 3$	$2^{12} \cdot 24 = 98\,304$

We end the discussion with this respectable achievement, although the story of combinatorial designs is far from finished. The interested reader can turn to Thompson (1983) and Conway and Sloane (1988) for a description of the groups which J. H. Conway found in connection with L. Conway denoted the automorphism group of L by $\cdot 0$ (pronounced "dotto"), and he defined $\cdot 1$ to be $\cdot 0$ divided by its center. It turns out that $\cdot 1$, along with two other automorphism groups, $\cdot 2$ and $\cdot 3$, are sporadic simple groups. The order of $\cdot 1$ is

$$2^{21} \cdot 3^9 \cdot 5^4 \cdot 7^2 \cdot 11 \cdot 13 \cdot 23.$$

Most simple groups belong to well-known families such as A_n, $PSL(n, q)$, or the groups of Lie type. However, 26 simple groups do not fall into these categories and are therefore called *sporadic simple groups*. Many of the 26 sporadic simple groups are associated with Aut L, including the largest, the *Monster*, a group of symmetries in a space of 196 884 dimensions. The Monster has order

$$2^{46} \cdot 3^{20} \cdot 5^9 \cdot 7^6 \cdot 11^2 \cdot 13^3 \cdot 17 \cdot 19 \cdot 23 \cdot 29 \cdot 31 \cdot 41 \cdot 47 \cdot 59 \cdot 71.$$

Notes

The relationship between the $(8, 16, 4)$ code of Exercise 9.4 and the lattice E_8, with highest contact number (240) in $\mathcal{R}^8$, is discussed in Conway and Sloane (1988).

The problem of finding the highest possible contact number for a lattice in $\mathcal{R}^k$ is very much unsolved. Conway and Sloane (1988) call this the *kissing number problem* and give a wealth of results. They also consider the related *packing problem, the covering problem* (in which one wants the least dense covering), and the so-called *quantizing problem* (which has application to data-compression). These subjects abound with open questions. For instance, although the obvious packing in $\mathcal{R}^3$ was proved by R. Hoppe in 1874 to have the highest possible contact number (see exercises), it has not been proved to be the densest possible packing.

In 1861 and 1873 E. Mathieu discovered five sporadic simple groups, M_{24}, M_{23}, M_{22}, M_{12}, and M_{11}, which are related to Steiner systems. The groups M_{24} and M_{23} are the automorphism groups of the $S(5, 8, 24)$ and $S(4, 7, 23)$ Steiner systems, respectively. The group M_{22} has index 2 in the automorphism group of the $S(3, 6, 22)$ Steiner system. The groups M_{12} and M_{11} are the automorphism groups of the $S(5, 6, 12)$ and $S(4, 5, 11)$ Steiner systems, respectively. These groups may also be defined in terms of permutations. For example, letting $x = (1\ 2\ 3\ 4\ 5\ 6\ 7\ 8\ 9\ 10\ 11)$, $y = (5\ 6\ 4\ 10)(11\ 8\ 3\ 7)$, and $z = (1\ 12)(2\ 11)(3\ 6)(4\ 8)(5\ 9)(7\ 10)$, we can write $M_{11} = \langle x, y \rangle$ and $M_{12} = \langle x, y, z \rangle$. It turns out that $|M_{11}| = 8 \cdot 9 \cdot 10 \cdot 11$ and $|M_{12}| = 8 \cdot 9 \cdot 10 \cdot 11 \cdot 12$. With 7920 elements, M_{11} is the smallest of the 26 sporadic simple groups.

Exercises

11.1 Show that the weight distribution of G_{23} is

weight	1	7	8	11	12	15	16	24
number	1	253	506	1128	1288	506	253	1.

11.2 Show that a Steiner system $S(5, 6, 12)$ can be constructed by fixing a codeword x of weight 12 in G_{24} and taking all codewords which intersect x in 6 places.

11.3 Show that M_{12}, the automorphism group of the above Steiner system, has $P(12, 5) = 12!/7!$ elements.

11.4 Show that the contact number of $(2\mathbb{Z})^n$ is $2n$.

11.5 What is the maximum possible contact number $c(X)$ for a lattice packing X in $\mathcal{R}^2$? How about in $\mathcal{R}^3$?

Bibliography

Aigner, M. (1979), *Combinatorial Theory*, Springer-Verlag, New York.

Alexanderson, G. L., Klosinski, L. F., and Larson, L. C. (1985), *The William Lowell Putnam Mathematical Competition, Problems and Solutions 1965–1984*, Mathematical Association of America, Washington, D.C.

Alon, N. and Spencer, J. (1992), *The Probabilistic Method*, Wiley, New York.

Anderson, I. (1974), *A First Course in Combinatorial Mathematics*, Oxford University Press, New York.

Anderson, I. (1987), *Combinatorics of Finite Sets*, Oxford University Press, New York.

Aschbacher, M. (1994), *Sporadic Groups*, Cambridge University Press, New York.

Batten, L. M. (1986), *Combinatorics of Finite Geometries*, Cambridge University Press, New York.

Biggs, N. (1993), *Algebraic Graph Theory*, Cambridge University Press, New York.

Bollobás, B. (1979), *Graph Theory*, Cambridge University Press, New York.

Bollobás, B. (1985), *Random Graphs*, Academic Press, London.

Bollobás, B. (1986), *Combinatorics: Set Systems, Hypergraphs, Families of Vectors, and Combinatorial Probability*, Cambridge University Press, New York.

Bollobás, B. (ed.) (1991), *Probabilistic Combinatorics and its Applications*, American Mathematical Society, Boston.

Cameron, P. J. (1994), *Combinatorics: Topics, Techniques, Algorithms*, Cambridge University Press, New York.

Cameron, P. J. and van Lint, J. H. (1991), *Designs, Graphs, Codes and Their Links*, Cambridge University Press, New York.

Comtet, L. (1974), *Advanced Combinatorics*, D. Reidel, Dordrecht.

Conway, J. H. (1986), *Atlas of Finite Groups, Maximal Subgroups, and Ordinary Characters for Simple Groups*, Oxford University Press, New York.

Conway, J. H. and Sloane, N. J. A. (1988), *Sphere Packings, Lattices, and Groups*, Springer-Verlag, New York.

Furstenberg, H. (1981), *Recurrence in Ergodic Theory and Combinatorial Number Theory*, Princeton University Press, Princeton.

Gessel, I. and Rota, G.-C., eds. (1987), *Classic Papers in Combinatorics*, Birkhäuser, Boston.

Gleason, A. M., Greenwood, R. E., and Kelly, L. M. (1980), *The William Lowell Putnam Mathematical Competition, Problems and Solutions 1938–1964*, Mathematical Association of America, Washington, D.C.

Graham, R. L. (1981), *Rudiments of Ramsey Theory*, American Mathematical Society, Providence.

Graham, R. L., Knuth, D. E., and Patashnik, O. (1994), *Concrete Mathematics*, Addison-Wesley, New York.

Graham, R. L., Rothschild, B., and Spencer, J. (1990), *Ramsey Theory*, 2nd edn, Wiley, New York.

Greene, D. H. and Knuth, D. E. (1990), *Mathematics for the Analysis of Algorithms*, Birkhäuser, New York.

Halberstam, H. and Roth, K. (1983), *Sequences*, Springer-Verlag, New York.

Hall, M. (1986), *Combinatorial Theory*, 2nd edn, Wiley, New York.

Harary, F. (1969), *Graph Theory*, Addison-Wesley, Reading.

Harary, F. and Palmer, E. M. (1973), *Graphical Enumeration*, Academic Press, New York.

Hardy, G. and Wright, E. (1979), *An Introduction to the Theory of Numbers*, Clarendon Press, Oxford.

Hartsfield, N. and Ringel, G. (1994), *Pearls in Graph Theory*, Academic Press, New York.

Herstein, I. N. (1996), *Abstract Algebra*, 3rd edn, Prentice-Hall, Upper Saddle River.

Hill, R. (1986), *A First Course in Coding Theory*, Clarendon Press, Oxford.

Johnson, D. L. (1990), *Presentations of Groups*, Cambridge University Press, New York.

Johnsonbaugh, R. (1993), *Discrete Mathematics*, Macmillan Publishing Company, New York.

Keedwell, A. D. (ed.) (1991), *Surveys in Combinatorics*, Cambridge University Press, New York.

Klee, V. and Wagon, S. (1991), *Old and New Unsolved Problems in Plane Geometry and Number Theory*, Mathematical Association of America, Ithaca.

Lyndon, R. (1986), *Groups and Geometry* (repr. with corrections), Cambridge University Press, New York.

McElicce, R. (1977), *The Theory of Information and Coding*, Addison-Wesley, Reading.

MacWilliams, F. J. and Sloane, N. J. A. (1978), *The Theory of Error-Correcting Codes*, North-Holland, Amsterdam.

Michaels, J. G. and Rosen, K. H. (1991), *Applications of Discrete Mathematics*, McGraw-Hill, New York.

Nesetril, J. *et al.* (eds.) (1990), *Mathematics of Ramsey Theory*, Springer-Verlag, New York.

Niven, I., Zuckerman, H., and Montgomery, H. (1991), *An Introduction to the Theory of Numbers*, 5th edn, Wiley, New York.

Palmer, E. (1985), *Graphical Evolution*, Wiley, New York.

Pless, V. (1989), *Introduction to the Theory of Error-correcting Codes*, Wiley, New York.

Pólya, G., Tarjan, E., and Woods, R. (1983), *Notes on Introductory Combinatorics*, Birkhäuser, Boston.

Roberts, F. S. (1984), *Applied Combinatorics*, Prentice-Hall, Englewood Cliffs.

Rotman, J. (1973), *The Theory of Groups: An Introduction*, 2nd edn, Allyn and Bacon, Boston.

Spencer, J. (1987), *Ten Lectures on the Probabilistic Method*, Society for Industrial and Applied Mathematics, Philadelphia.

Stanley, R. (1986), *Enumerative Combinatorics*, Wadsworth & Brooks/Cole Advanced Books & Software, Monterey.

Stanton, D. W. and White, O. E. (1986), *Constructive Combinatorics*, Springer-Verlag, New York.

Thompson, T. (1983), *From Error-correcting Codes Through Sphere Packings to Simple Groups*, Mathematical Association of America, Ithaca.

Tonchev, V. (1988), *Combinatorial Configurations: Designs, Codes, Graphs*, Wiley, New York.

Van Lint, J. H. and Wilson, R. M. (1992), *A Course in Combinatorics*, Cambridge University Press, New York.

Van Tilborg, H. C. A. (1988), *An Introduction to Cryptology*, Kluwer Academic Publishers, Norwell.

Wallis, W.D. (1988), *Combinatorial Designs*, Marcel Dekker, New York.

Welsh, D. (1988), *Codes and Cryptography*, Oxford University Press, New York.

Wilf, H.S. (1986), *Algorithms and Complexity*, Prentice-Hall, Englewood Cliffs.

Wilf, H. S. (1994), *Generatingfunctionology*, Academic Press, New York.

Index